Springer Theses

Recognizing Outstanding Ph.D. Research

Aims and Scope

The series "Springer Theses" brings together a selection of the very best Ph.D. theses from around the world and across the physical sciences. Nominated and endorsed by two recognized specialists, each published volume has been selected for its scientific excellence and the high impact of its contents for the pertinent field of research. For greater accessibility to non-specialists, the published versions include an extended introduction, as well as a foreword by the student's supervisor explaining the special relevance of the work for the field. As a whole, the series will provide a valuable resource both for newcomers to the research fields described, and for other scientists seeking detailed background information on special questions. Finally, it provides an accredited documentation of the valuable contributions made by today's younger generation of scientists.

Theses are accepted into the series by invited nomination only and must fulfill all of the following criteria

- They must be written in good English.
- The topic should fall within the confines of Chemistry, Physics, Earth Sciences, Engineering and related interdisciplinary fields such as Materials, Nanoscience, Chemical Engineering, Complex Systems and Biophysics.
- The work reported in the thesis must represent a significant scientific advance.
- If the thesis includes previously published material, permission to reproduce this must be gained from the respective copyright holder.
- They must have been examined and passed during the 12 months prior to nomination.
- Each thesis should include a foreword by the supervisor outlining the significance of its content.
- The theses should have a clearly defined structure including an introduction accessible to scientists not expert in that particular field.

More information about this series at http://www.springer.com/series/8790

Yue Zhu

Power System Loads and Power System Stability

Doctoral Thesis accepted by
The University of Manchester, Manchester, UK

Springer

Author
Dr. Yue Zhu
State Grid Corporation of China
Shanghai, China

Supervisor
Prof. Jovica V. Milanović
The Department of Electrical
and Electronic Engineering
The University of Manchester
Manchester, UK

ISSN 2190-5053 ISSN 2190-5061 (electronic)
Springer Theses
ISBN 978-3-030-37785-4 ISBN 978-3-030-37786-1 (eBook)
https://doi.org/10.1007/978-3-030-37786-1

This Springer imprint is published by the registered company Springer Nature Switzerland AG
The registered company address is: Gewerbestrasse 11, 6330 Cham, Switzerland

Supervisor's Foreword

The future power systems will be characterised by blurred boundaries between transmission and distribution system, by a mix of wide range of electricity-generating technologies (conventional hydro, thermal, nuclear and power electronics interfaced stochastic and intermittent renewable generation), responsive and highly flexible, typically power electronics interfaced, demand and storage with significant temporal and spatial uncertainty, proliferation of power electronics (HVDC, FACTS devices and new types of load devices) and significantly higher reliance on the use of measurement data including global (wide area monitoring) signals for system identification, characterisation and control and Information and Communication Technology (ICT) embedded within the power system network and its components. The reliance on wide area monitoring and embedded ICT coupled with the application of advanced data analytics methodologies will result in increased observability of the system. The proliferation of stochastic and intermittent power electronics interfaced generation technologies and loads, some of which still insufficiently understood and hence potentially less accurately modelled on the other hand, will result in increased uncertainty in system steady-state and dynamic behaviour and possibly inability to assess accurately, under some circumstances, some of the aspects of its performance. The increased uncertainty of system performance will pose an additional challenge to system controllability already aggravated by increased reliance on new stochastic, intermittent and generally less understood generation technologies and loads and ever reducing reliance on conventional generation. Considering the size of modern interconnected power systems, it is impractical and overly expensive to model every single system component in particular as some of them may not contribute significantly to overall system steady-state and dynamic performance.

This thesis, therefore, in the broadest terms, explores the extent to which one needs to develop accurate mathematical models of different system components when studying steady-state and dynamic behaviour of the system as a whole. It particularly focuses on the significance of accurate load modelling for power system stability studies. It develops a pioneering methodology and a concept for identifying critical loads and load model parameters in large power networks based on their influence on power system stability. The research described in the thesis first develops an automatic load modelling tool (ALMT) that can be used to automatically build a load model from measured power system data without human intervention. The ALMT can significantly increase the efficiency of load modelling and is suitable for online application and inclusion, as an additional advanced feature, in existing power quality monitors or fault level recorders. Secondly, it develops a pioneering framework based on Morris screening method for ranking power system load model parameters based on their influence on overall power system stability (voltage, frequency, transient and small disturbance angular stability) considering different load models and loading conditions. Ranking load model parameters helps determine the critical loads in the power system, and such only these critical loads need to be modelled accurately, thus reducing the resources needed for load modelling. Thirdly, the thesis proposes a novel probabilistic methodology for determining the required accuracy levels of critical load model parameters. The required accuracy level of load model parameters can be chosen based on the desired accuracy and confidence level of selected stability indices. This can both reduce the necessary investment in load modelling and improve the accuracy of, and confidence in, results of power system stability assessment.

The results of the research reported in the thesis have both, scientific and practical importance, and can be used as a guide for the development of practical load modelling programmes in industry. The thesis also represents an excellent basis for the continuation of the research both, in the aforementioned areas and in the more general area of identifying appropriate and sufficiently accurate mathematical models of other system components, generators, transformers, storage devices, etc., for the purpose of assessing global system steady-state or dynamic performance with desired accuracy and confidence level.

Manchester, UK
October 2019

Prof. Jovica V. Milanović
CEng, FIET, FIEEE
Deputy Head of Department

Abstract

Load models play an important role in power system stability analysis, and accurate load models are essential for and will benefit power system stability assessment. It is not necessary and practical though to model all loads very accurately, as load modelling requires significant financial and human resources. This thesis thus develops a methodology for identifying critical loads and load model parameters based on their influence on power system stability. Therefore, only those important loads and load model parameters would need to be modelled accurately, saving significant investments. The main outcome of this research is being able to identify the critical loads and load model parameters in the power system for different types of power system stability and to determine the required accuracy level of critical load model parameters so that the accuracy of power system stability analysis is not affected.

This research mainly contributes to the following areas. First, an automatic load modelling tool (ALMT) which was originally developed in the past, as part of the final year project, has been substantially improved and fine-tuned by adjusting software parameters for higher accuracy and implementing composite load model. The tool can be used to automatically build load model from measured power system data without human intervention. Second, a framework for ranking power system load model parameters for four types of power system stability has been developed. The framework is based on Morris screening method. Load model parameters are ranked according to their impact on power system stability indices. The ranking has been performed for four types of stability studies, three types of load models and four different loading conditions. Third, a methodology for determining the accuracy levels of critical load model parameters has been developed. The methodology is based on Monte Carlo simulation. Different uncertainty levels of load model parameters will result in various confidence levels of stability indices. The required accuracy level of load model parameters can be chosen according to the desired confidence level of stability indices. Fourth, the influence of stochastic dependence of load model parameters on the parameter ranking is investigated. Load model parameters have been obtained from actual field measurements. Based on them, the Gaussian copula is used to generate correlated load model parameters, which is then used for parameter ranking. Finally, the critical load locations in the power system are identified for different types of power system stability, and load buses are ranked irrespective of the load models used to represent the load of these buses.

Acknowledgements

Firstly, I would like to express my gratefulness to my supervisor Prof. Jovica V. Milanović, who constantly guided, supported and inspired me during my Ph.D. study. His attitude of keeping all the research at a high standard and focusing on details has stimulated me to dedicate myself to produce the highest quality work.

I will also express my gratitude to my great colleagues, especially Mr. Buyang Qi, Mr. Wentao Zhu, Dr. Atia Adrees, Dr. Kazi Hasan and Ms. Jelena Ponocko in the Power Quality and Power Systems Dynamics group of the School of Electrical and Electronic Engineering. It was constructive and inspiring having discussions with them throughout my research. The research cannot achieve what it has achieved today without their kind support in many aspects.

My special appreciation also goes to my dear friends Mr. Yuanpeng Zhou, Mr. Gaoyuan Liu and Mr. Yibin Peng. It is them who made my daily life joyful and exciting. They provided me with help and support whenever I need them.

Last but not least, I would like to express my thankfulness to my parents for their spiritual and financial support of my education and my daily life. Without their understanding and support, I will not be able to pursue a Ph.D. degree.

Contents

Nomenclature

Symbols

D	Generator damping constant
f	Frequency
H	Generator inertia constant
I	Current
P	Real power
Q	Reactive power
R	Resistance
S	Apparent power
s	Slip
T	Time constant
t	Time
V	Voltage
X	Reactance
Y	Admittance
Z	Impedance
α	Voltage exponent of the static exponential load model for the real power
β	Voltage exponent of the static exponential load model for the reactive power
Δ	Operator representing the variation
δ	Generator rotor angle
λ	Frequency ratio
μ	Mean value
σ	Standard deviation/eigenvalue
ω	Angular frequency

Subscriptions and Superscriptions

'	Referred value
0	Initial value
cr	Critical
d	Dynamic/d-axis
e	Electrical
G	Generator
L	Load
l	Leakage
m	Mechanical
n	Rated
q	q-axis
r	Rotor
s	Stator/Static
t	Transient

Acronyms

ALMT	Automated load modelling tool
BW	Butterworth
CCT	Channel Components Transform
CIGRE	International Council on Large Electric Systems
CPF	Continuous power flow
DFIG	Doubly fed induction generators
DG	Distributed generation
EAC	Equal-area criterion
EPRI	Electric Power Research Institute
FCC	Full converter connected
FDE	Kernel density estimation
FVSI	Fast voltage stability index
GA	Genetic algorithm
GB	Gigabyte
KCL	Kirchhoff's circuit laws
KVL	Kirchhoff's voltage law
LS	Least squares
LVRT	Low-voltage ride-through
MA	Moving average
MC	Monte Carlo
MISE	Mean integrated squared error
NETS-NYPS	New England Test System—New York Power System
ONR	Off-nominal turns ratio

OPF	Optimal power flow
PCC	Pearson's correlation coefficient
PSS	Power system stabiliser
PV	Power-voltage/photovoltaics
RES	Renewable energy sources
RMS	Root mean square
ROCOF	Rate of change of frequency
SA	Simulated annealing
SG	Savitzky-Golay
SNR	Signal-to-noise ratio
TRASI	Transient Rotor Angle Severity Index
TSI	Transient Stability Index
WECC	Western Electricity Coordinating Council

List of Figures

List of Tables

Chapter 1
Introduction

1.1 Power System Loads

Loads are one of major components of power systems. They range from household appliances to industrial motors. Load characteristics have a significant influence on system static and dynamic behaviour. If not appropriately represented and accounted for, it may result in suboptimal system performance, or even lead to system instability and collapse.

1.1.1 Load Characteristics

Load characteristics refer to the relationship between the real or reactive power and voltage and frequency. According to whether it is time-dependent or not, it can be divided into steady state and dynamic load characteristics. Load models are the mathematical representations of load characteristics. Load models represent the variation of active and reactive power as a function of voltage, frequency and time. Load models can also be classified into static load models and dynamic load models. The static load models represent the steady-state load characteristics, which is a function between the real or reactive power and the voltage and/or frequency. The dynamics load model represents the dynamic load characteristics, which is a function of active and reactive power as the dependent variable and voltage and/or frequency and time as the variable. More details of load models used in industry are given in Chap. 2.

Y. Zhu, *Power System Loads and Power System Stability*, Springer Theses,
https://doi.org/10.1007/978-3-030-37786-1_1

1.1.2 *Importance of Load Modelling*

Power system operation and control is significantly affected by load behaviours [1]. If load models can accurately capture the load behaviour during a disturbance, power system operators would be better informed about possible system behaviour and have more chance to successfully handle the emergency conditions and ultimately have a better control of the power system. Thus, in order to prevent the collapse and separation of power systems during disturbances and to ensure power system stability, sufficiently accurate load models should be used [2, 3].

Using inappropriate load models may result in unexpected disastrous results [4]. Load characteristics have significant influence on both steady state and dynamic behaviours of power systems. In order to accurately analyse power system behaviour and provide guidance to post-fault operations and planning of the power system, applying accurate load models in power system studies is necessary [5]. Many massive black-outs in history have showed the importance of accurate load models, including the Swedish blackout of 1983, the Tokyo grid collapse in 1987 [6], and the Western North America blackout in 1996 [7]. In response to events mentioned above, more research about load modelling had been carried out by some organisations. Most notably, the foundation of the Western Electricity Coordinating Council (WECC) in 2002 [8], the starting of a collaborative load modelling research program by Electric Power Research Institute (EPRI) in 2004 [9], and the establishment of an International Council on Large Electric Systems (CIGRE) load modelling working group [5].

Although significant amount of research about load modelling has been done, a large percentage of load models used nowadays are still those proposed in the 1990s [10], even though power systems have become much more complex and new types of loads have emerged since then, in particular, more non-linear loads connected through power electronic interface.

1.2 Power System Stability

Power System stability is the ability of a power system to operate at stable equilibrium point in normal conditions and to regain the stable operating point after a disturbance [5]. Power system stability is normally divided into following: small disturbance stability, transient stability, frequency stability, and voltage stability. In order to improve the efficiency and economics of power systems, they are operated more and more close to the stability limits. In the meantime, an increasing number of conventional generation sources have been and are being replaced by stochastic and less well understood renewable energy sources. Maintaining the power system stability therefore will face more risks due to the uncertainty of operating conditions brought by renewable generation [11].

1.2.1 Voltage Stability

Voltage stability refers to system being able to keep stable voltages at all buses in the power system in both normal operating condition and after the disturbances. A progressive drop or increase of voltage that could occur at some buses may lead to voltage collapse. The time period of interest for this type of stability is from few seconds to tens of minutes [12]. The power system is becoming more complex and the loading in general is increasing, together with economic and ecological concerns, voltage stability is becoming more and more crucial. In the past years, the problems associated with voltage stability have attracted more attention because many voltage collapse events have occurred in several countries, including Israel [13], Northwestern America [7], Italy [14], Greece [15], Arizona and Baja California [16], and India [17]. Therefore, maintaining voltage stability to prevent voltage collapse must be considered carefully when planning and operating modern power systems.

Voltage instability may take place when the load increases suddenly in a heavily loaded system [18], which is reflected in a sudden drop of system voltage. This voltage drop can be so quick that there may not be enough time for voltage control devices to take corrective actions to stop cascading power outage from happening [19]. To prevent such situation from happening, it is crucial for the power system designers and operators to ensure that the system is always operating at a safe distance from the voltage stability limit. Thus, one of the most important tasks of voltage stability studies is finding voltage stability limit and measuring the distance from it.

1.2.2 Frequency Stability

Frequency stability refers to the capacity of a power system to keep stable frequency after a major imbalance between generation and demand [12]. A 'lighter' power system, i.e., power system with low inertia can be more sensitive to frequency deviation and therefore faces more concerns about frequency stability. Frequency stability problems are usually related to inadequacies in equipment responses, improper coordination of control and protective devices or insufficient generation reserves. The examples can be found in [20–23].

The characteristic time frame of the frequency stability studies ranges from few seconds, which is the time frame of the responses of generator controls and protections, to several minutes, relating to the response of load voltage regulators. Thus, frequency stability can be both a short term phenomenon and a long term phenomenon. For example, the frequency instability caused by insufficient under frequency load shedding is a short term frequency stability problem [24], while the frequency stability problem caused by generation boiler and reactor protection and controls is a long term frequency stability problem [25].

1.2.3 Small Disturbance Stability

In this thesis, the small disturbance stability analysis focuses on small disturbance angular stability. Though in some other classifications [26], there is also a small disturbance voltage stability, it is not adopted in this thesis. The thesis focuses on using one stability index for each type of stability study for the purpose of quantitatively assessing the effects of load model parameters on them. Rotor angle stability refers to the ability of interconnected synchronous machines in the power system to remain in synchronism. The input mechanical torque equals the output electrical torque in a stable system so that all machines are operated at the same frequency. If there is a difference between the torques due to the system disturbance, then the rotor of one or more machines would deviate from the synchronous speed. This may result in rotor oscillations of growing amplitude due to lack of sufficient damping torque or a steady increase in rotor angle due to lack of sufficient synchronising torque [27]. In particular, small disturbance stability refers to the ability of the power system to maintain synchronism after being subjected to disturbances which are sufficiently small so that the system equations can be linearised [5]. Small disturbance includes load changes, increased power flow etc., which happens constantly in power systems. Since these occur constantly, if a practical power system is not small disturbance stable, it would not be physically operable.

The main problems associated with the small-disturbance stability in modern power systems is ensuring the electromechanical oscillations are well damped. The most often used small disturbance analysis method is the modal analysis of linear dynamic systems. In this method, the linearised differential equations are used to describe power system around an operating point, and to represent it using system state space model [26].

The power system is first modelled using vector-matrix form:

$$\dot{\boldsymbol{x}} = \boldsymbol{f}(\boldsymbol{x}, \boldsymbol{u}) \tag{1.1}$$

$$\boldsymbol{y} = \boldsymbol{g}(\boldsymbol{x}, \boldsymbol{u}) \tag{1.2}$$

In (1.1) and (1.2), $\boldsymbol{x}$ is a vector of n state variables, $\boldsymbol{u}$ is a vector of m system inputs, $\boldsymbol{y}$ represents a vector of p system outputs, and $\boldsymbol{f}$ and $\boldsymbol{g}$ are non-linear equations vectors. An equilibrium point at which $\boldsymbol{x} = \boldsymbol{x}_0$ and $\boldsymbol{u} = \boldsymbol{u}_0$ such that (1.1) is equal to zero is defined next. By making a small disturbance (Δ) at this point, Eq. (1.3) can be established

$$\dot{\boldsymbol{x}}_0 + \Delta\dot{\boldsymbol{x}} = \boldsymbol{f}(\boldsymbol{x}_0 + \Delta\boldsymbol{x}, \boldsymbol{u}_0 + \Delta\boldsymbol{u}) \tag{1.3}$$

Since only small disturbances are considered, a first order Taylor's series expansion of (1.3) can be used as a suitable approximation. The same can be done for Eq. (1.2). The linearised state space power system model in state space is then given by Eqs. (1.4) and (1.5).

$$\Delta\dot{\boldsymbol{x}} = \boldsymbol{A}\Delta\boldsymbol{x} + \boldsymbol{B}\Delta\boldsymbol{u} \tag{1.4}$$

$$\Delta\boldsymbol{y} = \boldsymbol{C}\Delta\boldsymbol{x} + \boldsymbol{D}\Delta\boldsymbol{u} \tag{1.5}$$

where $\boldsymbol{A}$ is a $n \times n$ state matrix, $\boldsymbol{B}$ is a $n \times m$ input matrix, $\boldsymbol{C}$ is a $p \times n$ output matrix, and $\boldsymbol{D}$ is a $p \times m$ feedthrough matrix. The expressions for them are:

$$\boldsymbol{A} = \begin{bmatrix} \frac{\partial f_1}{\partial x_1} & \cdots & \frac{\partial f_1}{\partial x_n} \\ \vdots & \ddots & \vdots \\ \frac{\partial f_n}{\partial x_1} & \cdots & \frac{\partial f_n}{\partial x_n} \end{bmatrix} \quad \boldsymbol{B} = \begin{bmatrix} \frac{\partial f_1}{\partial u_1} & \cdots & \frac{\partial f_1}{\partial u_m} \\ \vdots & \ddots & \vdots \\ \frac{\partial f_n}{\partial u_1} & \cdots & \frac{\partial f_n}{\partial u_m} \end{bmatrix}$$
$$\boldsymbol{C} = \begin{bmatrix} \frac{\partial g_1}{\partial x_1} & \cdots & \frac{\partial g_1}{\partial x_n} \\ \vdots & \ddots & \vdots \\ \frac{\partial g_p}{\partial x_1} & \cdots & \frac{\partial g_p}{\partial x_n} \end{bmatrix} \quad \boldsymbol{D} = \begin{bmatrix} \frac{\partial g_1}{\partial u_1} & \cdots & \frac{\partial g_1}{\partial u_m} \\ \vdots & \ddots & \vdots \\ \frac{\partial g_p}{\partial u_1} & \cdots & \frac{\partial g_p}{\partial u_m} \end{bmatrix}$$

The small-disturbance stability of a system is then determined by the eigenvalues of linearised system equations [28]. Therefore, by evaluating the eigenvalues of the system state matrix $\boldsymbol{A}$ of the linearised state space model of the power system, the small-disturbance stability of the power system can be determined.

1.2.4 Transient Stability

Transient stability is another type of rotor angle stability. It is the ability of the power system to maintain synchronism after a large disturbance. Influenced by the non-linear characteristics of the system, system responses would involve significant changes in bus voltages, power flows, generator rotor angle, and other system variables. In general, the loss of synchronism following a disturbance occurs within a very short time (3–5 s). The operating state after the disturbance is often different from the operating state before the disturbance. The type of disturbances studied here, e.g., short circuit faults, loss of major system component, etc., are much larger than in small disturbance studies, so that power system equations cannot be linearised [29].

A simple model consisting of a single generator connected to an infinite bus is used to demonstrate the elementary concept of transient stability, which is shown in Fig. 1.1. A reduced equivalent circuit of the system is illustrated in Fig. 1.2. The resistances are neglected.

Assuming that the mechanical power P_m is constant, the generator outputs electrical power P_e is given by (1.6) [26, 30, 31].

$$P_e = \frac{E'V}{X}\sin\delta \tag{1.6}$$

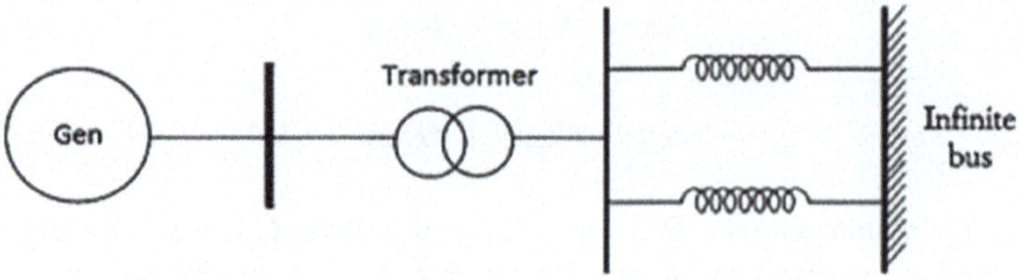

Fig. 1.1 One machine infinite bus system

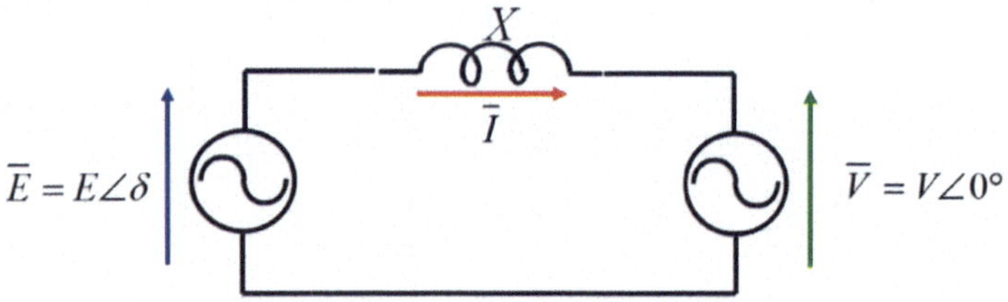

Fig. 1.2 Reduced equivalent circuit

where X is the equivalent network reactance, V refers to the voltage of infinite bus, and δ represents the generator rotor angle. The power-angle curve in Fig. 1.3

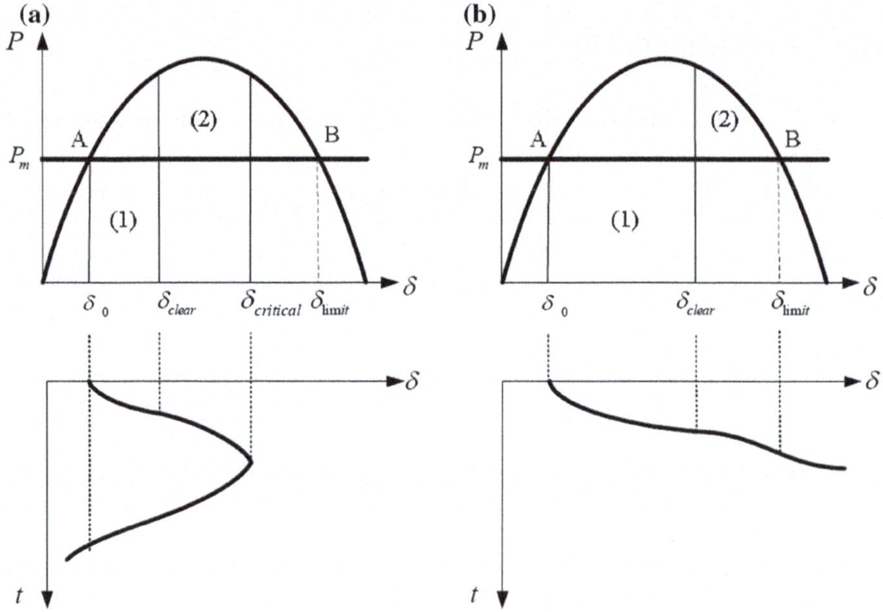

Fig. 1.3 Power-angle curve for one machine infinite bus system: **a** short clearing time, **b** long clearing time

shows the post-fault relationship between the P_e and δ. The pre-fault steady state operating point A has a rotor angle value of δ_0. Following a three-phase fault at the bus between the transformer and the transmission lines, the terminal voltage of the generator drops to 0 due to the fault, and the P_e becomes 0. Because there is no opposing torque from electrical power productions and the mechanical power is constant, the δ will increase.

When the δ is increased to δ_{clear}, the fault is cleared without disconnecting any line. Because the value of X is the same, the power-angle curve does not change. δ keeps increasing because of the kinetic energy gained during the fault, and reaches the critical value $\delta_{critical}$ when all kinetic energy has been dissipated. If $\delta_{critical}$ is smaller than the rotor angle value of the unstable point B, δ_{limit}, δ starts to decrease and will oscillate around the original operating point A with a decreasing amplitude due to the damping. This is illustrated in Fig. 1.3a. However, if δ increases over δ_{limit}, the system will be unstable and δ will keep increasing, which is illustrated in Fig. 1.3b.

The example above clearly describes the elementary concept of transient stability of a simple power system. The underlying principle is the ability of the power system to return to the original operating point or move from one steady state to another after large disturbances. The dynamic behaviour of generator after large disturbance is described by the swing curves, i.e., power or angle responses in time.

From the system theory point of view, transient stability is a strongly nonlinear, high-dimensional problem. In order to accurately analyse it, the numerical integration methods i.e. time-domain methods have to be used. The equal area criterion (EAC) is the simplest method for determining the transient stability. In Fig. 1.3, area (1) refers to the amount of kinetic energy the system gained during the fault, and area (2) represents the amount of energy that the system absorbs after the fault. In order for the system to reach the steady state and maintain synchronism, all the kinetic energy must be absorbed, which can be represented by equating the areas (1) and (2). In general EAC are strictly applicable to single machine infinite bus systems, though generalised EAC methods exist that could be applied to multi-machine cases [32].

The transient stability of a power system may be affected by many factors including:

- Post-fault transmission system impedance
- Fault clearing time
- Type of fault
- Fault location
- Generator parameters
- Load parameters.

Some of these factors are related to the operating condition of the system at the time when the disturbance occurs, whilst the others are related to the disturbance itself.

1.3 Review of the Past Work

1.3.1 Ranking of Power System Components for Small Disturbance Stability

Some previous work studied the ranking of power system components for different types of power system stability. In [33], residues and eigenvalue sensitivities were used to identify loads which have a significant influence on system damping. The power system loads were represented as a feedback of the power system transfer function. The magnitude of the deviation of transfer function eigenvalue describes the effects of load modelling on damping. If the deviation is large, then it means the load has a large influence on the mode. The deviation of eigenvalue is a function of both the residue and load sensitivity. Thus, it is affected by both the load and its location. In [34], a technique based on a quasi-optimisation procedure with the cost function reflecting shifts of selected eigenvalues along the real axis is applied to reveal loads having the biggest impact on damping of oscillations. The largest shifts of eigenvalues are obtained by all points of the cost function steepest descent trajectory in the space of unknown load parameters. Then the participation factors of load parameters are computed along the steepest descent trajectory, which is used to rank loads. Following that, the uniformly distributed sequences in hyperspace of load parameters is used to obtain a robust ranking of loads for small disturbance stability [35]. The ranking index is the sum of eigenvalue sensitivity with respect to all load model parameters of a load multiplied by the range of that parameter for all modes of interest. This index reflects the maximum possible variation of damping of critical modes caused by load model parameter uncertainty. The sensitivity factors are calculated at random points within the hyperspace of uncertain parameters.

In [36], the eigenvalue shift optimisation procedure is used to reveal the loads whose parameters have the largest effect on small disturbance stability. A ranking index is proposed which reflects the integral of squared sensitivity of selected eigenvalues with respect to an unknown load model parameter, computed along the trajectory that results in the maximum joint shift of eigenvalues along the real axis. More recently [37], evaluates nine sensitivity analysis techniques to identify the most influential parameters affecting power system small disturbance stability. The nine sensitivity techniques can be grouped into three categories: local, screening, and global sensitivity analysis. The accuracy, computational complexity and simulation time of these nine techniques have been compared. After comparison, the Morris screening method [38, 39] is found to be the best method, which provides a good balance between accuracy and efficiency.

1.3.2 Ranking of Power System Components for Voltage Stability

For voltage stability, in [40], two methods are proposed for identifying weak buses. One method is based on the relative bus voltage change between the initial operating point and the voltage collapse point. The second method is based on the sensitivity with respect to a vector specifying the direction of increase of the apparent power demand. The results produced by these two methods are very similar. Similarly, in [41], two methods of identifying weak buses in power systems are provided. The first method is based on the right singular vector [42], corresponding to a minimum singular value [43] of the power-flow Jacobian matrix, which indicated sensitive voltages. The second method is based on a voltage-collapse proximity indicator (VCIP), which identifies the weak buses and areas in the power system. The results obtained with these two methods are similar, and they are similar to results presented in [40]. However, the methods proposed in [41] have shorter simulation times.

Authors in [44] trained a Kohonen neural network [45] to rank buses in terms of voltage stability by using power flow analysis and the singular value decomposition method. The Kohonen network is one of unsupervised neural networks. It is very effective for pattern classification problems. Similarly, a Radial Basis Function (RBF) neural network [46] is trained to map the operating conditions of power systems to a voltage stability indicator and contingency severity indices corresponding to transmission lines [47]. The minimum singular value of Jacobian matrix is used as voltage stability indicator. A line stability index termed as fast voltage stability index (FVSI) is applied to identify the weakest bus in the system in [48]. The reactive power at a particular bus is increased until it reaches the instability point at bifurcation. The maximum loadability is the load connected at the particular bus at the instability point. The maximum loadability for each load bus is used for ranking buses. The bus that has the smallest value is ranked highest, which indicated that this bus has the lowest sustainable load. Similarly [49], applies FVSI to rank the line outage contingency. The ranking provides information that indicates the severity of the voltage stability condition in a power system due to line outage.

Paper [50] presents a fuzzy approach for ranking critical buses in a power system based on Line Flow Index (LFI) and voltage profiles at load buses. The Line Flow Index indicates the possible maximum load that can be connected to a bus in order to maintain stability before the system reaches its bifurcation point. Line Flow Index and voltage profiles at the load buses are represented in Fuzzy Set notation in this paper. Then they are evaluated using fuzzy rules to obtain Criticality Index. The buses are ranked based on this index. Paper [51] developed a load ranking scheme by a computational method based on the sensitivity of load margin to the real power variation of load. The real power of each load is changed individually, and the variation of load margin is obtained for each load, which is used to rank load buses. A modified equivalent reactive compensation method is proposed to rank loads according to the suitability of DG placement for compensating reactive power during occasions of reactive power shortage [52].

1.3.3 Ranking of Power System Components for Transient Stability

For transient stability, in [53], a proposed method uses the extended equal-area criterion to determine the unstable equilibrium point of the system. This direct stability criterion makes it possible to assess a convenient clearing time for computing the swing curves and hence to easily determine the system unstable equilibrium point; and to select the critical machines by observing them at the time corresponding to this unstable equilibrium point. In [54], the line potential energy method is used to derive a transient stability index which in turn is used for the identification of critical clusters of machines. Lines with small stability index are removed one by one until an island is detected, and the critical cluster and cutset are identified. In [55], a method is proposed for ranking synchronous generators in a power system, which is based on sensitivity analysis of electromechanical modes and takes into account the location of generators, their inertia, active and reactive power outputs, and control functions. The effect of a generator on the system is measured by its removal. The system integrity is checked through its simulation after the removal of each generator.

A technique that relies on a hybrid direct—time-domain method, called Single Machine Equivalent (SIME), is proposed to rank power system lines for transient stability [56]. SIME assesses stability by transforming the multi-machine power system into a one-machine infinite bus system. The stability margins are obtained using equal-area criterion, and used to rank contingencies on different lines. Another method of ranking power system lines is proposed in [57]. An index is presented based on the implicit expression of the stability region of the power system related to the controlling unstable equilibrium point. The index is estimated by the quadratic approximation of stability region. Finally, in [58], an transient stability index based on the kinetic energy derivative is proposed to rank power system lines. This index has a good correlation with the critical clearing time of the contingency and it allows comparing the severity of the contingencies themselves. This index can be calculated by simulating only the 'during fault' period in a time domain simulator, which makes it suitable for online application.

1.3.4 Summary

In summary, more research has been done for ranking power system loads for small disturbance stability than other types of stability studies. Methods that have been implemented include eigenvalue sensitivity, quasi optimisation, eigenvalue shift optimisation procedure, and uniformly distributed sequences. For voltage stability, though some studies have been done for load ranking, most of them rank all system buses. The sensitivity of load margin to the variation of load, FVSI and the modified equivalent reactive compensation method are applied for ranking loads. For transient stability, there were only methods for the identification of critical generators. The extended equal-area criterion, line potential energy, and sensitivity analysis of electromechanical modes are used for ranking generators. No studies have been done in frequency stability for ranking power system components. For voltage and small disturbance stability where the ranking of loads have been studied, only the critical loads are identified, but which load model parameters are more influential have not been looked into. In this research, the ranking of power system loads and their parameters are performed for all four types of power system stability studies.

1.4 Aims and Objectives of the Research

1.4.1 Aims of the Research

The key question that this project is trying to answer is: Do we need to develop load models (static and dynamic) for all buses in the network or only for a selected few that will make difference in terms of network dynamic performance. If we can develop methodology to identify important loads in the network and the level of their importance then resources can be devoted to focus on developing accurate load models for those buses only. The important loads can be identified and ranked based on their contribution for different types of stability studies, e.g., voltage stability, small disturbance stability, transient stability and frequency stability. Furthermore, not only that we could identify which loads are important but also how important they are (in terms of accuracy of determined model parameters). This may lead to answers like as long as the parameters of the load at bus 17, which has been identified as the most influential bus in the network from the point of view of voltage stability, are determined with 2% accuracy, the results of system studies will not be affected. This analysis could be done for a part of transmission or distribution network and for different types of stability studies considering all the uncertainties that load modelling may present in the future due to integration of distributed generations (DG) and mobile (electric vehicles) or stationary power electronics interfaced loads.

The principal aim of this research is to develop flexible simulation environment and generic framework for the analysis of the influence of power system loads on power system dynamic performance and to establish the level of required accuracy of load models for reliable assessment of power system dynamic performance including voltage stability, small disturbance stability, transient stability, and frequency stability. This research should be able to identify which loads and how accurately need to be modelled for different types of studies in a given network.

1.4.2 Objectives of the Research

In order to fulfil the aim of the research, the following objectives have to be met:

- Continue development, improve, and fine tune methodology and tool for load model development based on field measurements.
- Review and summarise state-of-the-art in the area of power system load modelling and to compile reported real life events of sub-optimal power system performance due to inadequate load modelling.
- Review and summarise state-of-the-art in the area of identification and ranking of critical power system parameters and components based on their influence on different aspects of power system stability.
- Develop methodology for identification and ranking of critical power system loads based on their influence on accuracy of assessment of four types of power system stability studies.
- Illustrate developed methodology on appropriate models of power system transmission and distribution network.
- Perform sensitivity analysis to identify the influence of variation of parameters of critical loads on accuracy of assessment of system stability and to correlate accuracy in load modelling with accuracy of system stability assessment.

1.5 Main Contributions of the Research

This research has contributed to several areas in the field of load modelling and load model parameter ranking. These contributions are summarised as follows. Paper numbers given after each paragraph indicate that the relevant contribution has been published in the international journal or in the proceedings of the international conference. A full list of author's thesis-based publications is provided in Appendix E.

1. **Development and improvement of the automated load modelling tool (ALMT). [E1]**
   Further development, improvement and tuning of the automated load modelling tool (ALMT) that can build load models from recorded power system data without human intervention. This tool, development of which started as undergraduate project, can automatically read the recorded power system data (voltage, current, active power, reactive power etc.). Then select the useful data that can be used for load modelling. After the useful data has been selected, they are filtered and modified, and appropriate load models are chosen according to the shape of selected responses. Following that, load model parameters are fitted using least squares methods. The fitted load model parameters are used to produce the simulated response, which is then compared with the recorded response. If they match each other, then the developed load model and corresponding parameters are recorded. If they do not match, the ALMT will roll back to load model selection stage and chose another load model. The ALMT can significantly increase the efficiency of load modelling and it is suitable for online application. The development and improvements made during the Ph.D. research were implementing composite

load model in ALMT, and tuning of the parameter values in different parts of the load modelling process in order to achieve higher accuracy of results.

2. **Ranking load model parameters by using the Morris screening method. [E2, E3, E4, E5, E6]**
Development of a method for ranking power system load model parameters and ranking load model parameters for all four types of power system stability studies based on their influence on relevant system stability. The load model parameter ranking process contains three stages. 28 ranking results in total have been obtained for the test network, which provides a comprehensive understanding of critical load model parameters. The main findings of ranking load model parameters are: 1. For critical load buses, all their parameters tend to be more critical than the same parameter for other buses. 2. Active load model parameters are usually more critical than reactive load model parameters. 3. Critical load locations generally remain unchanged for different stability studies and load models, though the ranking order may change. 4. The large loads are usually critical loads for all types of stability. Ranking load model parameters helps power system operators determine the critical loads in the power system, and only these critical loads need to be modelled accurately, thus many resources can be saved in the load modelling process.
3. **Determining the required accuracy levels of load model parameters of different loads by Monte Carlo method. [E2, E3]**
Development of a method for determining the required accuracy levels of load model parameters for four types of power system stability studies. The process also contains three stages. The first stage is generating uncertainties of load model parameters and renewable generation by Monte Carlo simulation. The second stage is performing power system stability analysis considering the uncertainties generated in Stage I. The third stage is identifying the confidence levels of each uncertainty level of critical load model parameters. In order to find the required accuracy level of critical load model parameter, one has to decide the desired confidence level of power system stability indices first, for example, 99% for voltage and frequency stability, 95% for small disturbance stability and 90% for transient stability. Then the corresponding accuracy levels can be found by looking at the confidence level—σ plots. This can both reduce the necessary investment in load modelling and improve the accuracy of power system stability assessment. The required accuracy levels of load model parameters obtained justify the ranking orders obtained by load model parameter ranking methods, that is, the more critical load model parameters require a higher confidence level. It was also found that for the same confidence level, the transient stability requires the highest accuracy level of load model parameters compared to other three types of stability.
4. **Investigating the influence of stochastic dependence of parameters on load model parameter ranking by Gaussian Copula method. [E3]**
In power systems, load model parameters are actually correlated with each other, and these correlations are nonlinear and non-Gaussian. In this research, the copula theory is applied to effectively model stochastic dependence among load model

parameters. The Pearson correlation coefficients between actual load model parameters have been obtained. The copula theory is then applied to model the correlation among these parameters. The load model parameter ranking has been performed again taking parameter correlation into consideration. The results are compared with the original ranking, unveiling the influence of stochastic dependence of parameters on load model parameter ranking. The comparison shows that the ranking order of load model parameters correlated with each other will be more similar than the ranking order when the correlation of these parameters is not considered.

5. **Identifying the critical load locations in power system for different types of system stability studies by Monte Carlo method. [E7]**
Identifying critical load locations in power system for four types of power system stability studies. In power systems, some load locations could be more influential on a certain type of power system stability than on the other, which is independent of the load type on that location The critical load locations are identified by calculating the Pearson correlation coefficients between the load variation of different locations and the variation of relevant power system stability indices for different types of stability studies. The research shows that for different types of power system stability studies, the critical load locations are different. For voltage stability, the important load locations are the load centres far from generators. For frequency stability, the critical load locations are load locations connected with tie lines. And for small disturbance and transient stability, the important load locations are those close to generators.

1.6 Thesis Overview

This thesis is divided into eight chapters. The first chapter is an introductory chapter. The main contents of the following seven chapters are summaries below.

Chapter 2—Power System Load Models and Load Modelling
This chapter can be divided into two parts. The first part is the literature review part. In the first part, different types of load models and load modelling methods have been reviewed and summarised. Six types of load models, including both static load models and dynamic load models, have been introduced and the model equations are given. Then two types of load modelling techniques—component based and measurement based load modelling methods have been introduced. The second part of chapter two describes the measurement based load modelling method. This part describes an automatic load model building tool (ALMT) whose development started as an undergraduate project and which was fully developed and finely tuned as a part of this research.

Chapter 3—Power System Stability Studies
This chapter discusses different power system stability indices used for the load

model parameter ranking. Firstly it introduces the power system stability indices used for four types of power system stability studies in this research. Following that it provides a critical review and comparison of different types of stability indices.

Chapter 4—Probabilistic Assessment and Sensitivity Analysis in Stability Studies

This chapter discusses the overall approach used for load model parameter ranking comprising three essential parts, namely, probabilistic assessment, sensitivity analysis, and stochastic dependence. The chapter discusses each of them and how they were implemented. More specifically, the probabilistic assessment method is used for determining the required accuracy levels of different load model parameters, the sensitivity analysis techniques are used for ranking power system load model parameters, and the stochastic dependence is used to investigate the influence of correlation between load model parameters.

Chapter 5—Load Model Parameter Ranking for Different Types of Power System Stability Studies

This chapter discusses the main part of this research, which is identifying critical load model parameters according to their influence of power system stabilities. First, it introduces the test network for the load model parameter ranking and explains the procedure for parameter ranking. Then the load model parameter ranking is carried out for three types of load models and four types of power system stability using Morris screening method. The parameter ranking is performed considering various loading conditions. The results of the ranking are illustrated using heatmaps. Finally, a summary of the parameter ranking is made at the end of this chapter.

Chapter 6—Factors Affecting Load Model Parameter Ranking

In this chapter, various factors that may affect the importance of load model parameters have been investigated. These include load size, stochastic dependence among load model parameters, load model type and load location. The chapter describes how each of these factors affects the established importance of load model parameters. Finally, for load locations specifically, it tries to identify critical load locations for each type of power system stability irrespective of the specific load models on these locations.

Chapter 7—Required Accuracy Level of Critical Load Model Parameters

After identifying the important load model parameters for power system stability studies, the next question is how accurately these parameters need to be modelled. This chapter solves this problem by obtaining the accuracy levels required for different load model parameters in order to achieve certain confidence levels of chosen power system stability indices. The accuracy levels of chosen critical load model parameters have been obtained for different load models, different power system stability studies and different loading conditions. The relationships between the variation of load model parameters and power system stability indices are obtained by running Monte Carlo simulations. This chapter thus provides guidance for determining the required accuracy of load models for individual loads in power system based on their influence on different types of power system stability studies.

Chapter 8—Conclusions and Future Work
This final chapter summaries the whole thesis and provides possible directions for future works.

References

1. Han D, Ma J, He R-M, Dong Z-Y (2009) A real application of measurement-based load modeling in large-scale power grids and its validation. IEEE Trans Power Syst 24:1756–1764
2. Craven R, Michael M (1983) Load characteristic measurements and representation of loads in the dynamic simulation of the Queensland power system. In: 1983 CIGRE and IFAC symposium
3. Renmu H, Jin M, Hill DJ (2006) Composite load modeling via measurement approach. IEEE Trans Power Syst 21:663–672
4. Price W, Chiang H-D, Clark H, Concordia C, Lee D, Hsu J, Ihara S, King C, Lin C, Mansour Y (1993) Load representation for dynamic performance analysis. IEEE Trans Power Syst 8:472–482
5. Milanovic JV, Gaikwad A, Borghetti A, Djokic SZ, Dong ZY, Andrew Halley, Korunovic LM, Villanueva SN, Ma J, Pourbeik P, Resende F, Sterpu S, Villella F, Yamashita K, Auer O, Karoui K, Kosterev D, Leung SK, Mtolo D, Zali SM, Collin A, Xu Y (2014) Modelling and aggregation of loads in flexible power networks. J. M. CIGRE WG C4.605
6. Ohno T, Imai S (2006) The 1987 Tokyo blackout. In: 2006 IEEE PES power systems conference and exposition, pp 314–318
7. Kosterev DN, Taylor CW, Mittelstadt WA (1999) Model validation for the August 10, 1996 WSCC system outage. IEEE Trans Power Syst 14:967–979
8. W. M. a. V. W. Group, WECC MVWG Load Model Report ver. 1.0. In: WECC, 2012
9. D. S. a. W. P. S. Ranade, Advanced load modeling—entergy pilot study. In: EPRI2004
10. Collin AJ (2013) Advanced load modelling for power system studies. Insitute for Energy Systems, University of Edinburgh, Edinburgh, UK
11. Preece R (2013) A probabilistic approach to improving the stability of meshed power networks with embedded HVDC lines. Ph.D., School of Electrical and Electronic Engineering, The University of Manchester, Manchester, UK
12. Kundur P, Paserba J, Ajjarapu V, Andersson G, Bose A, Canizares C, Hatziargyriou N, Hill D, Stankovic A, Taylor C, Van Cutsem T, Vittal V (2004) Definition and classification of power system stability IEEE/CIGRE joint task force on stability terms and definitions. IEEE Trans on Power Syst 19:1387–1401
13. Hain Y, Schweitzer I (1997) Analysis of the power blackout of June 8, 1995 in the Israel Electric Corporation. IEEE Trans Power Syst 12:1752–1758
14. Corsi S, Sabelli C (2004) General blackout in Italy Sunday September 28, 2003, h. 03: 28: 00. IEEE Power Eng Soc Gener Meet 2004:1691–1702
15. Vournas C (2004) Technical summary on the Athens and Southern Greece blackout of July 12, 2004. National Technical University of Athens
16. Ferc N (2012) Arizona-southern california outages on 8 September 2011: causes and recommendations. FERC and NERC
17. Romero JJ (2012) Blackouts illuminate India's power problems. IEEE Spectr 49
18. Suzuki M, Wada S, Sato M, Asano T, Kudo Y (1992) Newly developed voltage security monitoring system. IEEE Trans on Power Syst 7:965–973

19. Suganyadevia MV, Babulal CK Estimating of loadability margin of a power system by comparing voltage stability indices. In: 2009 international conference on control, automation, communication and energy conservation. INCACEC, 2009, pp 1–4
20. O'Sullivan J, Rogers A, Flynn D, Smith P, Mullane A, O'Malley M (2014) Studying the maximum instantaneous non-synchronous generation in an island system—frequency stability challenges in Ireland. IEEE Trans Power Syst 29:2943–2951
21. Zhao C, Topcu U, Li N, Low S (2014) Design and stability of load-side primary frequency control in power systems. IEEE Trans Autom Control 59:1177–1189
22. Ali E, Abd-Elazim S (2011) Bacteria foraging optimization algorithm based load frequency controller for interconnected power system. Int J Electr Power Energy Syst 33:633–638
23. Jiang L, Yao W, Wu Q, Wen J, Cheng S (2012) Delay-dependent stability for load frequency control with constant and time-varying delays. IEEE Trans Power Syst 27:932
24. C. T. F. Rep., Analysis and modeling needs of power systems under major frequency disturbances. January 1999
25. Younkins T, deMello F, Dunlop R, Fenton F, Intrabartola J, Kundur P, Littman B, Russell T (1983) Guidelines for enhancing power plant response to partial load rejections. IEEE Trans Power Appar Syst (United States) 102
26. Kundur P (1994) Power system stability and control, 2nd edn, vol 1. McGraw-Hill, New York
27. Preece R (2013) Improving the stability of meshed power networks: a probabilistic approach using embedded HVDC lines. Springer Science & Business Media
28. Liapunov AM (2016) Stability of motion. Elsevier
29. Xu Y (2015) Probabilistic estimation and prediction of the dynamic response of the demand at bulk supply points. Ph.D. School of Electrical and Electronic Engineering, University of Manchester, Manchester
30. Grainger JJ, Stevenson WD, Chang GW (1994) Power system analysis. vol 621. McGraw-Hill New York
31. Machowski J, Bialek J, Bumby JR, Bumby J (1997) Power system dynamics and stability. Wiley
32. Chiodo E, Lauria D (1994) Transient stability evaluation of multimachine power systems: a probabilistic approach based upon the extended equal area criterion. IEE Proc-Gener, Transm Distrib 141:545–553
33. Hiskens I, Milanovic J (1997) Locating dynamic loads which significantly influence damping. IEEE Trans Power Syst 12:255–261
34. Makarov YV, Maslennikov VA, Hill DJ (1996) Revealing loads having the biggest influence on power system small disturbance stability. IEEE Trans Power Syst 11:2018–2023
35. Maslennikov VA, Milanovic JV, Ustinov SM (2002) Robust ranking of loads by using sensitivity factors and limited number of points from a hyperspace of uncertain parameters. IEEE Trans Power Syst 17:565–570
36. Makarov YV, Hill DJ, Milanovic JV (1997) Effect of load uncertainty on small disturbance stability margins in open-access power systems. In: Proceedings of the thirtieth Hawaii international conference on system sciences, 1997, pp 648–657
37. Hasan KN, Preece R, Milanović JV (2017) Priority ranking of critical uncertainties affecting small-disturbance stability using sensitivity analysis techniques. IEEE Trans Power Syst 32:2629–2639
38. Iooss B, Lemaître P (2015) A review on global sensitivity analysis methods. In: Uncertainty management in simulation-optimization of complex systems, Springer, pp 101–122
39. King D, Perera B (2013) Morris method of sensitivity analysis applied to assess the importance of input variables on urban water supply yield–a case study. J Hydrol 477:17–32
40. Obadina O, Berg G (1990) Identifying electrically weak and strong segments of a power system from a voltage stability viewpoint. In IEEE proceedings C (generation, transmission and distribution), pp 205–212
41. Chen Y-L, Chang C-W, Liu C-C (1995) Efficient methods for identifying weak nodes in electrical power networks. In: IEEE proceedings-generation, transmission and distribution, vol 142, pp 317–322

42. Demmel JW (1997) Applied numerical linear algebra vol 56. Siam
43. Lof PA, Smed T, Andersson G, Hill DJ (1992) Fast calculation of a voltage stability index. IEEE Trans on Power Syst 7:54–64
44. Song Y, Wan H, Johns A (1997) Kohonen neural network based approach to voltage weak buses/areas identification. In: IEEE proceedings-generation, transmission and distribution, vol 144, pp 340–344
45. Lo K, Peng L, Macqueen J, Ekwue A, Cheng D (1995) Kohonen neural network as an aid tool in power system security operation
46. Billings SA, Zheng GL (1995) Radial basis function network configuration using genetic algorithms. Neural Netw 8:877–890
47. Wan H, Ekwue A (2000) Artificial neural network based contingency ranking method for voltage collapse. Int J Electr Power Energy Syst 22:349–354
48. Musirin I, Rahman TA (2002) Estimating maximum loadability for weak bus identification using FVSI. IEEE Power Eng Rev 22:50–52
49. Musirin I, Rahman TA On-line voltage stability based contingency ranking using fast voltage stability index (FVSI). In: Transmission and distribution conference and exhibition 2002: Asia Pacific. IEEE/PES 2002, pp 1118–1123
50. Shankar S, Ananthapadmanabha DT (2011) Fuzzy approach to critical bus ranking under normal and line outage contingencies. arXiv preprint arXiv:1103.0127
51. Li Y, Chiang H-D, Li H, Chen Y-T, Huang D-H, Lauby MG Power system load ranking for voltage stability analysis. In: Power engineering society general meeting. IEEE, 2006, p 8
52. Ettehadi M, Ghasemi H, Vaez-Zadeh S (2013) Voltage stability-based DG placement in distribution networks. IEEE Trans Power Deliv 28:171–178
53. Xue Y, Pavella M (1993) Critical-cluster identification in transient stability studies. In: IEEE proceedings C (generation, transmission and distribution), pp 481–489
54. Yuan W, Chan K, Zhang Y Identification of critical cluster in transient stability study using line potential energy method. In: Power engineering society general meeting, 2006. IEEE, p 6
55. Alhasawi FB, Milanovic JV (2012) Ranking the importance of synchronous generators for renewable energy integration. IEEE Trans Power Syst 27:416–423
56. Ruiz-Vega D, Ernst D, Ferreira CM, Pavella M, Hirsch P, Sobajic D (2000) A contingency filtering, ranking and assessment technique for on-line transient stability studies. In: International conference on electric utility deregulation and restructuring and power technologies, DRPT 2000, pp 459–464
57. Xue A, Shen C, Mei S, Ni Y, Wu FF, Lu Q A new transient stability index of power systems based on theory of stability region and its applications. In: Power engineering society general meeting, 2006. IEEE, p 6
58. Grillo S, Massucco S, Pitto A, Silvestro F Indices for fast contingency ranking in large electric power systems. In: MELECON 2010–2010 15th IEEE Mediterranean electrotechnical conference, 2010, pp 660–666

Chapter 2
Power System Load Models and Load Modelling

Load modelling has been long recognised as one of the most important parts of power system modelling. Most of the currently used load models were developed many years ago, and after the significant changes in load structure and characteristics over the years, they are now to a larger extent inappropriate [1]. Although the importance of accurate load models for power system studies has been emphasised by the power system research community, the industry still apply typical static load models.

The steady state and dynamic performance of power system are heavily affected by the load characteristics [2, 3]. Therefore, in order to analyse power systems properly, accurate load models, as well as proper models of generators, transmission lines, and transformers are necessary. Load modelling however is a complicated procedure, because many factors need to be considered simultaneously, including information on the load structure and the appropriate method for the assessment and validation of the load model needs to be chosen.

2.1 Load Models

In this section, some of the most common load models are presented. The existing load models can be divided into two groups—the static load models and dynamic load models. The static load models include: static exponential, polynomial, linear, and static induction motor load model. The dynamic load models include: exponential dynamic, dynamic induction motor, and composite load model. The word 'static' and 'dynamic' refers to the type of studies these models are suitable for. The static studies are about the steady state analysis of the system, hence static load models would suffice, while the dynamic studies are about the transient analysis of the system, where modelling of loads using differential equations is required to capture variation of the response with time.

Y. Zhu, *Power System Loads and Power System Stability*, Springer Theses,
https://doi.org/10.1007/978-3-030-37786-1_2

2.1.1 Static Exponential Load Model

The static exponential load model is one of the most frequently used load models around the world. It is described by (2.1) and (2.2):

$$P = P_0\left(\frac{V}{V_n}\right)^{k_{pu}}\left(\frac{f}{f_n}\right)^{k_{pf}} \tag{2.1}$$

$$Q = Q_0\left(\frac{V}{V_n}\right)^{k_{qu}}\left(\frac{f}{f_n}\right)^{k_{qf}} \tag{2.2}$$

where P and Q are active and reactive power of the load at voltage V and frequency f. P_0 and Q_0 are the active and reactive power drawn by the load when it is connected to the rated voltage V_n and the rated frequency f_n. The parameters k_{pu} and k_{qu} are used to describe the load variation with respect to voltage change, and k_{pf} and k_{qf} are used to describe the load variation due to frequency change.

The load change due to frequency variation can be neglected and usually is because it is less frequent and significant than the load change caused by voltage variation. Therefore, the static exponential load model can be simplified as (2.3) and (2.4):

$$P = P_0\left(\frac{V}{V_n}\right)^{k_{pu}} \tag{2.3}$$

$$Q = Q_0\left(\frac{V}{V_n}\right)^{k_{qu}} \tag{2.4}$$

The parameter k_{pu} and k_{qu} represent the sensitivity of active and reactive power with respect to voltage variation [4]. The loads are called constant power, constant current and constant impedance load model if the exponential parameters in (2.3) and (2.4) are set to 0, 1, and 2. These load models are all among most frequently used load models. For example, many resistive loads like heaters and hot plates are normally modelled as constant impedance load [5, 6].

2.1.2 Polynomial Load Model

Beside the static exponential load model, another frequently used static load model is polynomial load model. The voltage dependent form of the polynomial load model is as follows:

$$P = P_0\left[p_1\left(\frac{V}{V_n}\right)^2 + p_2\left(\frac{V}{V_n}\right) + p_3\right] \tag{2.5}$$

$$Q = Q_0\left[q_1\left(\frac{V}{V_n}\right)^2 + q_2\left(\frac{V}{V_n}\right) + q_3\right] \tag{2.6}$$

This load model is also referred to as the ZIP load model, because it contains constant impedance (Z), constant current (I) and constant power (P) components. Parameters p_1 and q_1 stand for the percentage of the constant impedance load, p_2 and q_2 the percentage of the constant current load, and p_3 and q_3 the percentage of the constant power load in the total load mix. The sum of parameters p_1, p_2, and p_3, and q_1, q_2, and q_3 should be equal to 1.

2.1.3 Linear Load Model

The linear load model is often used in the small disturbance stability studies where the voltage does not have a large variation. The linear load model is expressed as (2.7) and (2.8):

$$P = P_0\left(a_0 + a_1\frac{V}{V_n}\right) \tag{2.7}$$

$$Q = Q_0\left(b_0 + b_1\frac{V}{V_n}\right) \tag{2.8}$$

There are typically two parameters a_0 and a_1 for modelling P, and two parameters b_0 and b_1 for modelling Q [7]. However, it is found that the reactive power behaviour is more closely following the ZIP load model, therefore it is more accurate to model active power using linear load model and reactive power using ZIP load model [7].

2.1.4 Static Induction Motor Load Model

The static induction motor load model is often used to represent the induction motor load. The model is derived from the induction motor equivalent circuit as illustrated in Fig. 2.1.

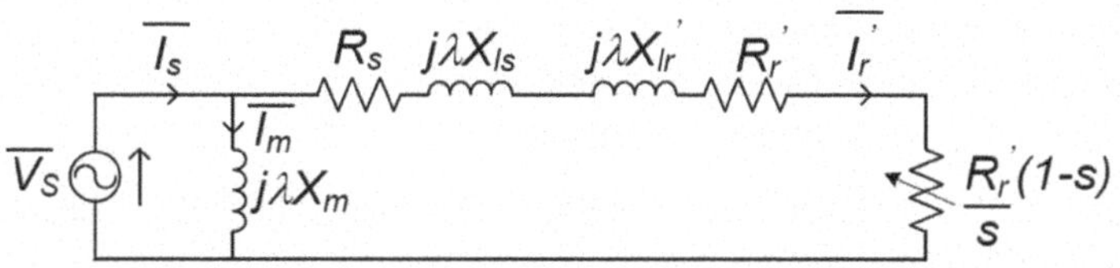

Fig. 2.1 Induction motor equivalent circuit

The meanings of the symbols in Fig. 2.1 are:

R_s Stator resistance
$R_r{}'$ Referred rotor resistance
X_{ls} Stator leakage reactance
$X_{lr}{}'$ Referred rotor leakage reactance
X_m Magnetising reactance
s Slip
λ Frequency ratio.

The active and reactive powers of the induction motor are given by (2.9) and (2.10):

$$P = \left(R_s + \frac{R_r}{s}\right)\frac{V^2}{\left(R_s + \frac{R_r}{s}\right)^2 + (X_{ls} + X_{lr})^2} \tag{2.9}$$

$$Q = (X_{ls} + X_{lr})\frac{V^2}{\left(R_s + \frac{R_r}{s}\right)^2 + (X_{ls} + X_{lr})^2} + \frac{V^2}{X_s} \tag{2.10}$$

2.1.5 Exponential Dynamic Load Model

The exponential dynamic load model is used to represent the exponential power recovery following a voltage step change [8]. It is described by (2.11) and (2.12):

$$T_p\frac{dP_r}{dt} + P_r = P_s(V) - P_t(V) = P_0\left(\frac{V}{V_0}\right)^{\alpha_s} - P_0\left(\frac{V}{V_0}\right)^{\alpha_t} \tag{2.11}$$

$$P_l = P_r + P_0\left(\frac{V}{V_0}\right)^{\alpha_t} \tag{2.12}$$

where:

V_0 Initial voltage value
T_p Real power recovery time constant
P_r Real power recovery
P_0 Initial value of real power
P_l Real power drawn by the load
α_s Steady state real power voltage parameter
α_t Transient real power voltage parameter.

The shape of the response after a voltage disturbance and physical meaning of parameters are illustrated in Fig. 2.2.

The reactive power response can be described using the same form as active power, with changes in corresponding symbols and parameters. The model can accurately

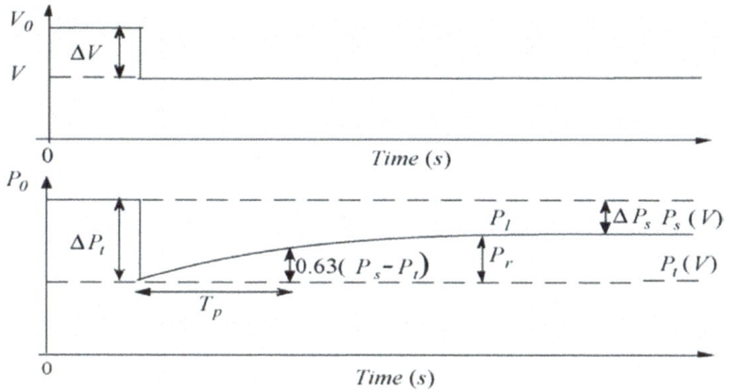

Fig. 2.2 Exponential dynamic model response

describe long term load response following limited voltage variation of participation of induction motors [9]. However, it is generally not sufficiently accurate to be used for short term load responses.

2.1.6 Composite Load Model

Being the most advanced load model, the composite load model is widely used in dynamic studies [1, 10]. The obtained power response is valid for most power system stability studies. The structure of composite load model has many variations, the most often used one is an induction motor model connected with a ZIP load model in parallel [1, 11–15], as shown in Fig. 2.3.

The current i flowing through the composite load model consists of i_s and i_d. i_s refers to the current flowing in the static part of the load, and i_d represents the current flowing through the dynamic part of the load, as shown by (2.13).

$$i = i_s + i_d \tag{2.13}$$

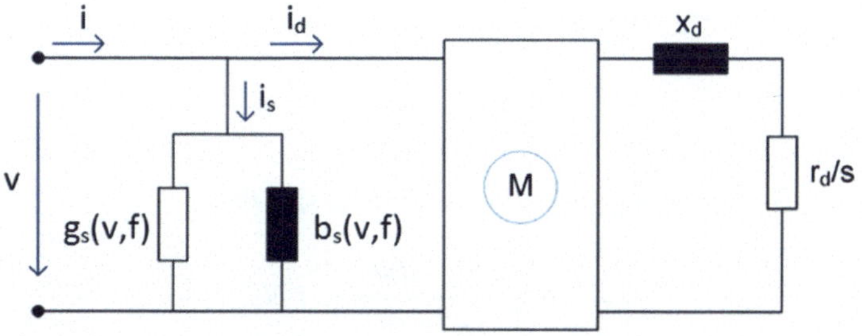

Fig. 2.3 The composite load model

The static part is modelled as a polynomial load model as described by (2.5) and (2.6). The static current is expressed by (2.14):

$$i_s = v\left(g_s \frac{p}{p_0} + jb_s \frac{q}{q_0}\right)\frac{v_0^2}{v^2} \tag{2.14}$$

where:

$$g_s = \frac{p_0}{v_0^2}(1 - t_{m0}) \tag{2.15}$$

$$b_s = -\frac{q_0 - p_0 t_{m0} \frac{s_0}{s_{cr}}}{v_0^2} \tag{2.16}$$

where s_0 is the normal operating slip, s_{cr} is the critical slip, t_{m0} represents the percentage of dynamic load, v is the voltage of the composite load model, g_s refers to the static conductance, and b_s is the static susceptance. The dynamic part of the load is modelled as a simplified induction motor model, and the current flowing through this part of the load is given by (2.17):

$$i_d = \frac{v}{z_d} = \frac{v}{\frac{r_d}{s_0} + jx_d} \tag{2.17}$$

where:

$$x_d = \frac{v_0^2 s_0 s_{cr}}{p t_{m0}\left(s_0^2 + s_{cr}^2\right)} \tag{2.18}$$

$$r_d = x_d s_{cr} \tag{2.19}$$

where x_d is dynamic reactance and r_d represents dynamic resistance. In transient stability studies, the current of the dynamic part of the load is expressed by (2.20):

$$i_d = \frac{\Psi_s - \Psi_r}{x_d} \tag{2.20}$$

where the stator flux Ψ_s is expressed by:

$$\Psi_s = -j\frac{v}{f_e} \tag{2.21}$$

In addition, a rotor voltage equation need to be satisfied:

$$r_d i_d + \frac{1}{2\pi F_{nom}} \frac{d\Psi_r}{dt} = js\Psi_r \tag{2.22}$$

Finally the slip is calculated by

$$s = f_e - \frac{\omega_r}{2\pi} \tag{2.23}$$

where f_e is the electrical frequency, and the derivative of the rotor speed by:

$$\frac{d\omega_r}{dt} = \frac{T_e - T_m}{T_j} \tag{2.24}$$

where T_m is mechanical torque, T_e is electrical torque and T_j represents the acceleration time constant. From (2.13) to (2.24), the parameters s_0, s_{cr}, t_{m0}, and T_j are the dynamic load model parameters that need to be determined.

2.2 Load Modelling Methodology

This section summarise the two major load modelling methodologies—component based approach and measurement based approach, with their corresponding advantages and disadvantages clearly stated. The component based approach is a bottom-up approach, while the measurement based approach is a top-down approach [16]. In general, if a load model contains multiple model components, for example, the ZIP load model and the induction motor load model in parallel, both the component based approach and measurement based approach can be used. If the load consists of single component, usually the measurement based approach is used.

2.2.1 *Component Based Load Modelling Approach*

The component based approach is a bottom-up modelling methodology. By using this approach, the load models are obtained from the following aspects: historical data of load classes connected to a substation, the structure and composition of load components in each load class, and individual characteristics of every load components. A typical load model structure and an illustrative set of parameter values provided by the component based approach are shown in Fig. 2.4.

According to Fig. 2.4, based on load consumption, the load can be divided into different load classes, which are agricultural, industrial, commercial, and residential. It is more difficult to represent load for some load classes than for others, because the load is distributed and a lot of data is required. The step following identifying load classes is to determine the load components and their corresponding percentage contribution within each load class. There are typical load components for each load class, which represents the majority of the end user power consumption. The common load components are lighting, refrigeration, cooling, and heating. It is important to determine how much each of the components contributes to that load class. Customer

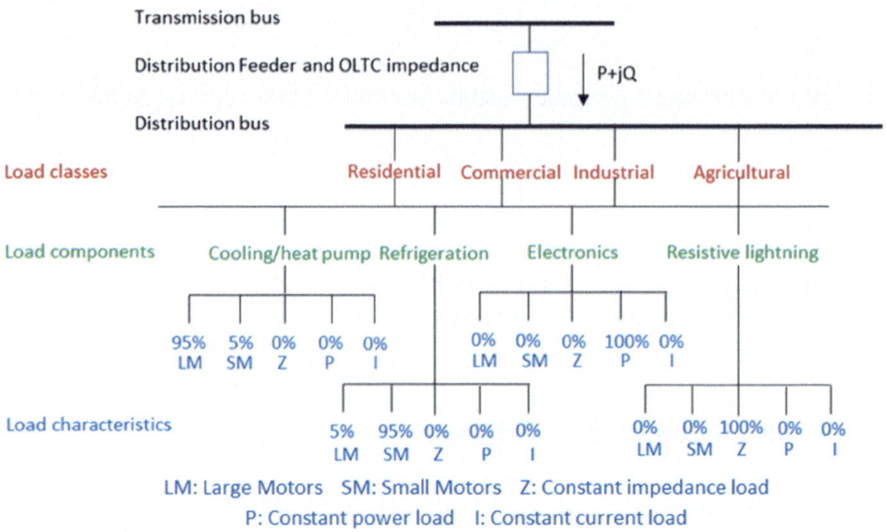

Fig. 2.4 Component based load modelling approach example [17]

surveys are often used in order to fulfil this task and this is typically complicated and time-consuming exercise.

Another approach is representing each load components by a number of different models of actual electrical circuits based on their electrical characteristics. In Fig. 2.4, LM stands for large electric motor model, SM represents small electric motor model, Z means constant impedance load model, P refers to constant power load model, I represents constant current load model. This approach can provide a more detailed load model because models of actual electrical circuits can represent load voltage responses more accurately. It is also important to remember that load characteristics may contain static and dynamic models. For example, cooling is often modelled as large induction motor, while lightning is usually modelled as constant impedance loads.

The advantages of component based load modelling approach are summarised as follows:

- It uses the generally available load class data from individual substations.
- It is suitable for developing composite load models.
- If the structure and composition of load components are known or can be acquired by surveys, field measurements are not required.
- It can be applied to different systems and conditions.
- The assessment of system performance sensitivity to variation in load composition variation using simulation studies is reasonably straightforward.

The disadvantages of component based load modelling approach are:

- The obtained load models normally assume that there is no variation in load characteristics, structure, and composition at different times.

- If the load model structure information is obtained for one substation, another substation may not have the same load model structure.
- It is difficult to identify load model parameters if a new or undefined load type is connected to the system.
- The load parameters can vary over a wide range depending on age, the manufacturer, and end-use application even if the fraction of each load component in load mix is identical.
- It is difficult for transmission system operators to use the component based load modelling approach, as the ownership and location of load devices in customer facilities may not be directly accessible.

2.2.2 Measurement Based Load Modelling Approach

The measurement based approach is a top-down method. It uses the recorded power system events and disturbances to develop the load models. The load model structures must be assumed first in order to develop load models and corresponding parameters. The load model parameters are obtained by using curve fitting and parameter identification methods from the measured data of real power systems. Advanced parameter identification techniques, for example, artificial intelligence are needed in order to develop parameters for complex load models. The identification procedure includes developing an appropriate mathematical function and corresponding parameters so that the function can reproduce the load responses after a disturbance. The identification is done by relating the recorded change in voltage and active and reactive power of the load. Figure 2.5 illustrates a flow chart of the load modelling approach.

The first step of measurement based load modelling approach is collecting power system data of voltage, current, and active and reactive power in time domain. These data are usually collected from the secondary side of a medium-voltage transformer. In the next step, some data processing techniques, like Moving Average and Savitzky-Golay filtering techniques are used to remove the high frequency noises from the data [18]. In the third step, a suitable load model structure is selected for fitting load model parameters. Following that, some optimisation techniques are applied, such as Least-Squares Algorithm, Genetic Algorithm, and Simulated Annealing, in order to minimise the difference between the model response and measured response [19–21]. When using these optimisation techniques, the user must choose an initial value, which could have a significant influence on the optimisation results. For the fifth step, the load model responses are compared with the recorded response by simulation tools to validate the derived load model and corresponding parameters. If the model responses do not match the recorded responses, then the process should roll back to step three. Finally, if the model responses match the recorded responses, then the derived load model and corresponding parameters are accepted and will be applied in future simulations.

The benefits of measurement based load modelling approach include:

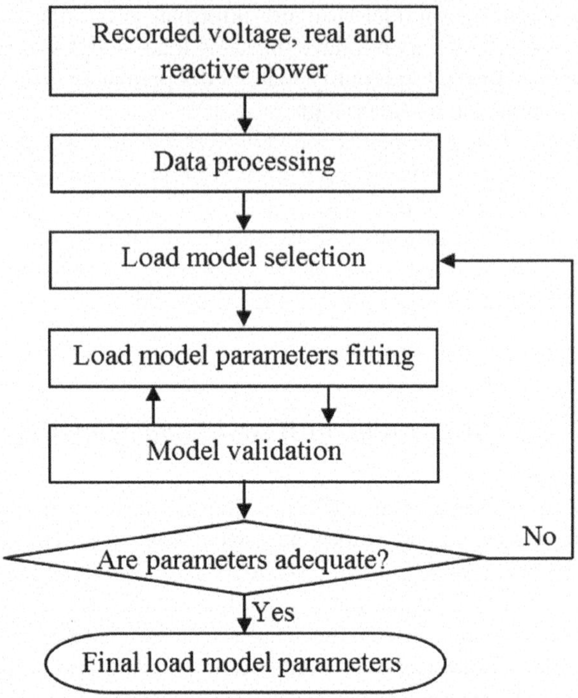

Fig. 2.5 Flow chart of measurement based load modelling approach

- The method is simpler than the component based load modelling approach for that the directly recorded load responses form the real power system are used.
- Temporal changes in load can be captured if competent measurements are recorded for a long enough period of time.
- This method can be applied to any load.

The drawbacks of measurement based load modelling approach:

- If the large disturbance data is not recorded, then the accurate load model cannot be obtained. However, large disturbances in power system cannot be applied intentionally, thus the measurement may take a very long period of time in order to record enough data for parameter identification.
- For a composite load model, there may be multiple solutions giving different parameter values after the optimisation process.
- The load variation that is not associated with voltage or frequency change should not be used for load model parameter identification. Otherwise, the derived load model parameters will be inaccurate.

2.3 Automatic Identification of Power System Load Models

As a part of this research, an Automated Load Modelling Tool (ALMT) is developed in order to automate the load modelling process [22]. The ALMT can automatically derive load models and corresponding parameters from measured power system responses. It is able to identify and derive load model parameters for polynomial, exponential dynamic and composite load model. The Load model parameter identification process can be divided into three stages: data processing, load model selection, and load model parameter fitting.

2.3.1 Data Processing

The first stage of ALMT is data processing. The recorded power system response cannot be directly used to derive load model parameters. They need to be processed first. The data is automatically imported into ALMT as long as the location where the data is stored is given. The ALMT can read data from .xlsx and .csv files. The recorded data set consists of three phase currents, three phase voltages, real power, reactive power, and frequency. The phase voltages are converted into symmetrical components using (2.25) because they are usually unbalanced. The positive sequence of the symmetrical voltage components is used as V in the following processes.

$$V_s = A^{-1} V_p \tag{2.25}$$

where V_s is voltage represented using symmetrical components and V_p is voltage represented using phase components.

Not all recorded data can be used to develop load models in measurement based load modelling. The most important requirement is that the recorded voltage must have significant voltage change. The transformer tap change often results in a voltage change between 0.5 and 2.5% [23, 24]. The P and Q change due to the voltage change should also fall in a reasonable range. Therefore, the next step is the identification of the data containing significant voltage variation. These data points are identified by comparing adjacent data values. The qualified data not only need to have a large magnitude change, but also need to last long enough to prevent the consecutive voltage changes being regarded as one voltage change. The flow chart for identifying events is illustrated in Fig. 2.6.

The next step is using an appropriate filter to filter the selected data sets, because the recorded power system data always contain some noise induced by data acquisition devices, and the data used in case studies are the originally recorded raw data that have not been pre-filtered. The Savitzky-Golay (SG) [25], Moving Average (MA) [18], and Butterworth (BW) [25] filter, which are widely used for data filtering, are compared with each other in order to choose the most appropriate one. A SG filter is a digital filter that can be applied to a set of digital data points for the purpose of

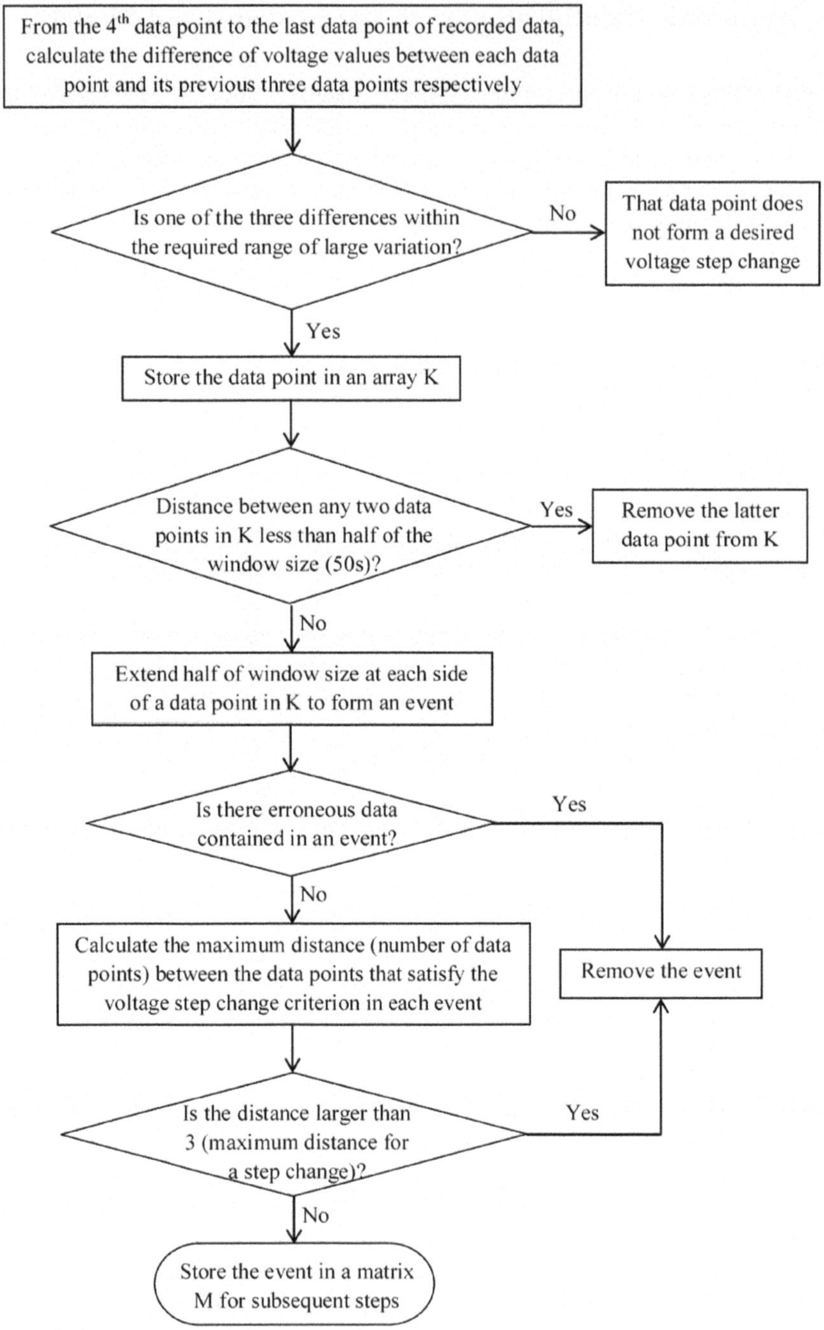

Fig. 2.6 The flow chart for selecting events

smoothing the data. This is achieved by convolution. By fitting successive sub-sets of adjacent data points with a low-degree polynomial by the method of linear least squares. When the data points are equally spaced an analytical solution to the least-squares equations can be found, in the form of a single set of convolution coefficients that can be applied to all data sub-sets, to give estimates of the smoothed signal at the central point of each sub-set. It can increase the Signal to Noise Ratio (SNR) and keep the useful signal undistorted.

MA filter performs well in reducing Gaussian white noise and keep the sharp step responses at the same time. MA filter uses the average value of data points around a data point to replace its original value. This can be described by (2.26):

$$y_{AV}(i) = \frac{1}{2N+1}(y(i+N) + y(i+N-1) + \cdots + y(i-N)) \tag{2.26}$$

where $y_{AV}(i)$ is the smoothed value for the ith data point, N is the number of neighbouring data points on either side of ith data point, and $2N+1$ is the span.

BW filter is a low-pass filter with the feature of having maximum level of flat magnitude response in the pass-band. The transfer function of BW filter in z-transform domain when implemented in MATLAB is given by (2.27):

$$H(z) = \frac{b_1 + b_2 z^{-1} + \cdots + b_{n+1} z^{-n}}{1 + a_2 z^{-1} + \cdots + a_{n+1} z^{-n}} \tag{2.27}$$

where $a_2, a_3, \ldots, a_{n+1}$ are denominator coefficients and $b_1, b_2, \ldots, b_{n+1}$ are numerator coefficients. The filtering results of recorded V, P, and Q are illustrated in Figs. 2.7, 2.8 and 2.9.

The blue original signal curves are obtained by connecting the digital data points recorded at 10 Hz. Three criteria have been used for the determination of the most suitable filter, which are efficiency in nose removing, preserving dynamic feature

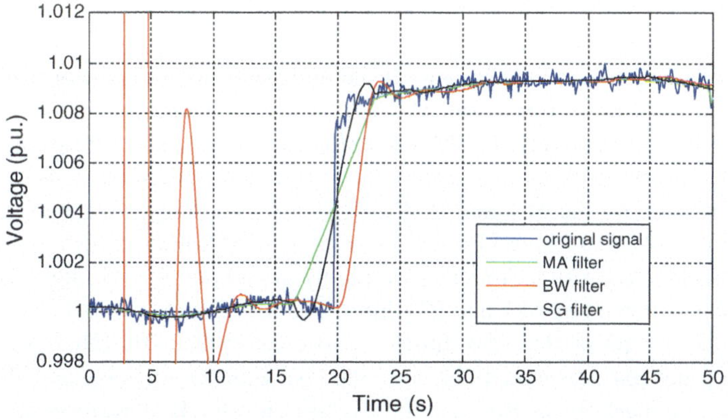

Fig. 2.7 Three filters apply on voltage signal

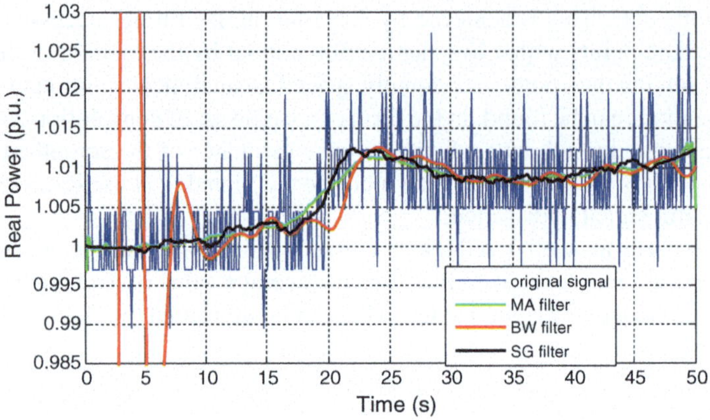

Fig. 2.8 Three filters apply on active power signal

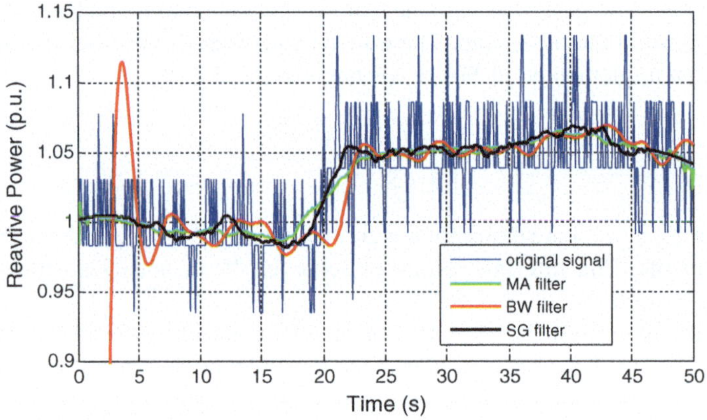

Fig. 2.9 Three filters apply on reactive power signal

of the original signal, and the quality at the end (start and the end of disturbance) points. In terms of noise removing, MA filter is the best, and SG and BW filter are second and third respectively. The MA filter has the smoothest results of three filters. In terms of preserving dynamic feature of the original signal, SG filter is performing the best, followed by the MA and BW filter. There is a delay after the step change at 20 s of the responses filtered by BW filter, and the MA filtered responses become too slow for the step change. In terms of quality at end points, SG filter has the best performance, followed by MA and BW filter. The MA filter has no effect on the start point and the end point. The BW filter in particular suffers from the issue that the initial data point starts from zero, because the BW filter have a delay and all samples before the time 0 are zeros. Because the SG filter has the best performance for two criteria and is second for the other criterion, it is chosen as the filter used in ALMT.

Table 2.1 Three types of possible load response shapes

Response shape	Response name
	Static response
	First order dynamic recovery response
	High order dynamic oscillatory response

After the data are filtered, they are converted into per unit value for further processing. The rated voltage at the load bus and the initial steady state power are used as base values. Following per unit value conversion, the recorded responses have to be divided into three categories based on the response shapes. Three types of P and Q responses caused by voltage change are identified as static response, first order dynamic recovery response, and high order dynamic oscillatory response, as illustrated in Table 2.1. The ALMT can automatically match each response with these three types of response shapes. This is realised by identification and comparison of the shape features of these responses. The flow chart of the process is shown in Fig. 2.10.

The final step of data processing is data modification. The overall quality of measured responses has been improved a lot after previous steps since noise is removed and only the data sets that can be used to build load models upon are stored for the following stage. However, some fluctuations still exist in filtered responses. Some of these fluctuations are results of remaining noise, and some may be caused by the natural fluctuation of load. Moreover, there are some signal distortions caused by the filter when the filter 'slows down' the response. For example, at the voltage step change point, the effect of filter results in a ramp instead of a sharp change. This delay in response needs to be removed before the next stage. The modification is based on the average value before and after the voltage step change, the filtered response, and the time of the step change. Figures 2.11, 2.12 and 2.13 shows the original response, filtered response, and modified response for three types of response shapes.

2.3.2 Load Model Selection

Three types of load models have been integrated into ALMT, which are polynomial load model, exponential dynamic load model and composite load model. Details of these load models are presented in Sect. 2.1. After the recorded responses are divided into different types, they can be used to develop corresponding load models. The ALMT will choose polynomial load model for static responses, exponential dynamic load model for first order recovery response, and composite load model for

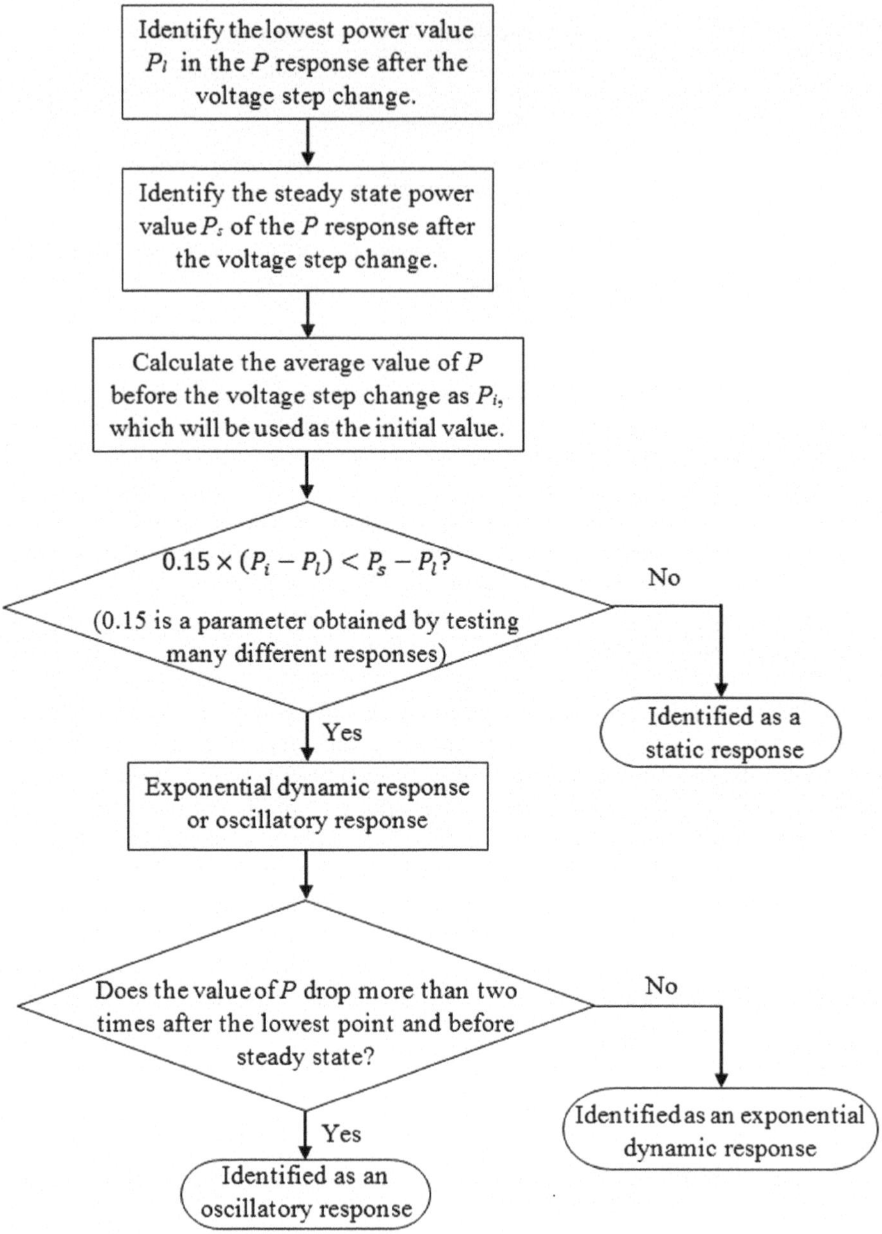

Fig. 2.10 Flow chart of the process for response differentiation

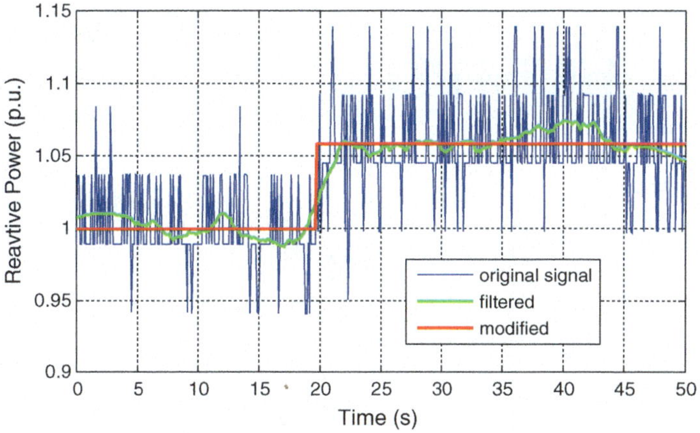

Fig. 2.11 Processing a static response

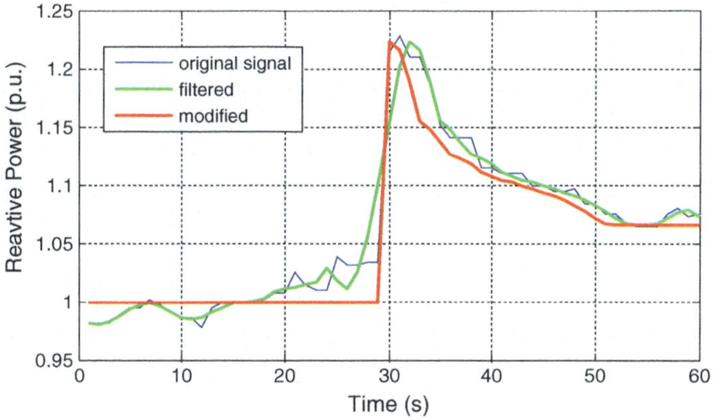

Fig. 2.12 Processing a first order recovery response

high order oscillatory response. Only one load model should be used to represent load in a certain time slot. If there are different types of responses recorded in a time slot, then the load model is chosen according to the dominant type of responses. After the load model is chosen and its parameters are fitted, ALMT will simulate the model response to see whether it can match the measured load response. If the simulated response and recorded response does not match with each other, then the load model will be switched to another one and the process will be repeated until the simulated and measured load response can match with each other.

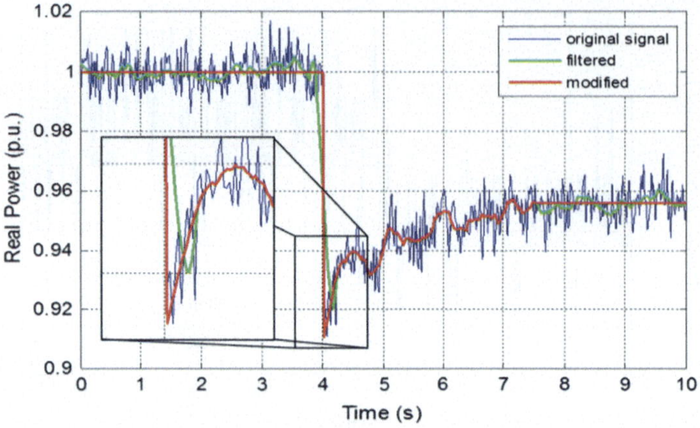

Fig. 2.13 Processing a high order oscillatory response

2.3.3 Load Model Parameter Fitting

Three parameter fitting algorithms, i.e. Least Squares Method (LS) [26], Genetic Algorithm (GA) [27], and Simulated Annealing (SA) [28] are tested within MATLAB, using sample data to compare their performance. The index used to determine the accuracy of developed load model and its parameters is the mean square errors between the recorded response and simulated response. The error for the real power is expressed by (2.27):

$$P_{error} = \frac{1}{n}\sum_{i=1}^{n}\left(P_{model_i} - P_{measured_i}\right)^2 \tag{2.27}$$

where P_{model_i} and $P_{measured_i}$ refers to the ith simulated and recorded real power values and n represents the number of measured events taken into account. The parameters obtained using three algorithms are given in Tables 2.2, 2.3 and 2.4 for eight recorded events. It takes an average 0.84 s to process each event.

According to Tables 2.2, 2.3 and 2.4, the least squares method has the best performance based on the average error and the variances of parameters. SA has smaller average errors than GA, but its variance is large. In theory, the GA and SA should have better performance than LS since they are more advanced methods. However, in ALMT, the maximum iteration number used for GA and SA is not very large for the sake of parameter identification speed, as ALMT may have to fit parameters for hundreds of recorded events in order to obtain a final load model parameter set for a time slot. If GA and SA are set for a high accuracy, the process of parameter identification would become computationally very expensive and thus impractical. The accuracy of three methods is compared in Table 2.5 with varying computation time. It can be seen that in order to achieve similar parameter variance as with the

Table 2.2 ZIP parameters obtained by least squares method

Event	p_1	p_2	p_3	P_{error}	q_1	q_2	q_3	Q_{error}
1	0.510	0.401	0.090	5.65×10^{-5}	2.676	0.406	−2.082	1.75×10^{-5}
2	0.593	0.399	0.007	2.03×10^{-5}	2.128	0.396	−1.523	2.15×10^{-5}
3	1.061	0.402	−0.462	2.37×10^{-5}	2.788	0.406	−2.194	2.76×10^{-5}
4	0.849	0.401	−0.251	6.36×10^{-5}	2.453	0.405	−1.858	3.62×10^{-5}
5	0.938	0.401	−0.339	1.19×10^{-5}	2.741	0.405	−2.146	4.87×10^{-5}
6	0.199	0.400	0.401	7.65×10^{-5}	2.195	0.405	−1.600	4.05×10^{-5}
7	0.361	0.400	0.239	5.27×10^{-5}	2.313	0.406	−1.719	5.03×10^{-5}
8	1.155	0.402	−0.556	5.61×10^{-5}	3.280	0.406	−2.685	3.64×10^{-5}
Average	0.708	0.401	−0.109	4.52×10^{-5}	2.572	0.404	−1.976	3.48×10^{-5}
Variance	0.118	0.000	0.119	–	0.144	0.000	0.145	–

Table 2.3 ZIP parameters obtained by genetic algorithm

Event	p_1	p_2	p_3	P_{error}	q_1	q_2	q_3	Q_{error}
1	0.770	0.054	0.171	8.94×10^{-5}	0.345	0.151	0.496	4.35×10^{-4}
2	0.337	−0.109	0.777	3.42×10^{-5}	0.273	0.451	0.295	2.22×10^{-4}
3	0.473	0.328	0.196	4.29×10^{-5}	0.813	0.057	0.114	2.58×10^{-4}
4	−0.562	1.868	−0.309	3.03×10^{-5}	3.259	−0.287	−1.972	5.41×10^{-4}
5	0.131	0.985	−0.110	1.07×10^{-5}	0.356	0.505	0.135	3.94×10^{-4}
6	0.954	0.298	−0.245	3.61×10^{-5}	0.430	1.315	−0.752	1.45×10^{-4}
7	0.608	0.275	0.115	2.93×10^{-5}	0.991	0.391	−0.377	4.59×10^{-4}
8	0.385	0.045	0.565	3.32×10^{-5}	0.660	0.867	−0.543	2.07×10^{-4}
Average	0.387	0.468	0.145	3.83×10^{-5}	0.891	0.431	−0.215	2.72×10^{-4}
Variance	0.213	0.428	0.143	–	0.980	0.245	0.757	–

Table 2.4 ZIP parameters obtained by simulated annealing

Event	p_1	p_2	p_3	P_{error}	q_1	q_2	q_3	Q_{error}
1	−2.341	5.682	−2.342	5.83×10^{-5}	8.169	−10.84	3.675	1.80×10^{-5}
2	−1.091	3.656	−1.565	2.03×10^{-5}	4.825	−5.339	1.515	2.23×10^{-5}
3	−0.173	2.148	−0.977	2.94×10^{-5}	9.020	−10.99	2.974	3.92×10^{-5}
4	1.230	0.669	−0.895	2.02×10^{-5}	2.986	−0.352	−1.633	3.86×10^{-5}
5	−0.930	4.177	−2.246	1.21×10^{-5}	9.507	−10.02	1.520	1.46×10^{-5}
6	−2.671	5.822	−2.152	8.95×10^{-5}	14.34	−20.00	6.672	2.55×10^{-5}
7	−1.209	2.529	−0.32	4.18×10^{-5}	3.104	−4.345	2.227	2.43×10^{-5}
8	0.181	3.455	−2.633	1.47×10^{-5}	6.709	−5.115	−0.590	5.18×10^{-5}
Average	−0.875	3.517	−1.641	2.57×10^{-5}	7.332	−8.375	2.045	1.02×10^{-4}
Variance	1.654	3.060	0.693	–	14.41	35.89	6.570	–

Table 2.5 Parameter variance obtained with different optimisation techniques as computational time increases

Computation time [s]	GA			SA			LS		
	p_1	p_2	p_3	p_1	p_2	p_3	p_1	p_2	p_3
1	0.204	0.416	0.138	1.586	2.954	0.676	0.118	0.000	0.119
10	0.122	0.143	0.121	0.137	0.652	0.285	0.112	0.000	0.115
100	0.107	0.000	0.115	0.112	0.002	0.117	0.107	0.000	0.113
1000	0.106	0.000	0.113	0.107	0.000	0.113	0.106	0.000	0.113

least squares method, GA and SA should have 10 times longer computation time. Thus, LS method is implemented in the software because of its efficiency.

2.3.4 *Case Studies and Results*

In order to illustrate the performance of ALMT, two case studies have been carried out. In Case I, the recording sampling rate is 1 Hz, which is suitable for identifying static load model parameters and exponential dynamic load model parameters. In Case II, the sampling rate is 50 Hz, which is high enough for building composite load models. Because high order oscillatory responses can only be adequately captured and subsequently reproduced if the sampling rate is higher than 16.67/20 ms for 60/50 Hz system. Tables 2.6 and 2.7 are examples of Case I data and Case II data respectively.

The Case I consists of 3 GB recorded load responses of 86,400 samples per day over a period of 5 months at 15 11 kV distribution substations. The data includes three phase currents, three phase voltages, real power, and reactive power for each substation. The load composition at a substation is different at different time periods. Thus it is necessary to build different load models for different time slots. The load models for weekdays and weekends are different, and during weekdays, daytime and night will have their own load models. The Case II contains data recorded at 50 Hz sampling rate of a single substation. There are only several events available for building load model. Three phase voltage, active power, and reactive power are included in the data. Tables 2.8, 2.9 and 2.10 show obtained load model parameters for some buses.

Table 2.6 Case I data example

Timestamp	V_1 [V]	V_2 [V]	V_3 [V]	I_1 [A]	I_2 [A]	I_3 [A]	P [MW]	Q [Mvar]
07:00:00	6536	6553	6579	276.51	264.11	268.64	2.982	0.697
07:00:01	6535	6554	6578	276.93	264.61	268.44	2.985	0.697
07:00:02	6535	6554	6578	276.93	264.61	268.44	2.985	0.697
07:00:03	6535	6555	6578	276.50	265.21	268.81	2.988	0.695
07:00:04	6534	6557	6577	276.84	265.18	269.50	2.991	0.694
07:00:05	6533	6556	6576	276.84	264.63	269.22	2.988	0.694
07:00:06	6533	6556	6576	276.84	264.63	269.22	2.988	0.694
07:00:07	6534	6557	6578	276.22	263.54	268.31	2.979	0.692
07:00:08	6532	6557	6577	275.49	261.90	266.78	2.965	0.686
07:00:09	6533	6558	6577	276.73	262.00	267.54	2.973	0.687
07:00:10	6530	6558	6575	277.51	261.90	268.18	2.977	0.688

Table 2.7 Case II data example

Timestamp UTC	P [MW]	Q [MVAR]	V_1 [V]	V_2 [V]	V_3 [V]
16:44:32.100	3.464	0.101	6564	6594	6580
16:44:32.120	3.461	0.100	6563	6592	6579
16:44:32.140	3.461	0.098	6562	6592	6578
16:44:32.160	3.462	0.098	6562	6592	6578
16:44:32.180	3.462	0.098	6560	6592	6578
16:44:32.200	3.464	0.098	6560	6591	6578
16:44:32.220	3.464	0.098	6560	6589	6577
16:44:32.240	3.467	0.097	6559	6590	6575
16:44:32.260	3.462	0.097	6558	6588	6576
16:44:32.280	3.471	0.098	6558	6587	6576
16:44:32.300	3.464	0.098	6558	6588	6575
16:44:32.320	3.470	0.097	6557	6588	6575
16:44:32.340	3.474	0.099	6558	6588	6575
16:44:32.360	3.471	0.101	6558	6589	6576
16:44:32.380	3.472	0.101	6559	6589	6576
16:44:32.400	3.480	0.101	6559	6590	6577

Table 2.8 Load model parameters of Substation 6, Case I

Parameters		p_1	p_2	p_3	q_1	q_2	q_3
Weekday daytime	Average	1.284	0.398	−0.694	3.046	0.399	−2.470
	Most probable	1.253	0.382	−0.655	3.122	0.412	−2.463
	Min.	1.203	0.374	−0.744	2.917	0.369	−2.558
	Max.	1.382	0.421	−0.635	3.223	0.435	−2.359
Weekday night	Average	0.759	0.400	−0.159	2.799	0.401	−2.200
	Most probable	0.738	0.411	−0.145	2.807	0.398	−2.188
	Min.	0.717	0.382	−0.223	2.674	0.377	−2.316
	Max.	0.807	0.411	−0.099	2.955	0.424	−2.084
Weekends	Average	0.732	0.400	−0.132	2.814	0.402	−2.216
	Most probable	0.749	0.388	−0.147	2.806	0.400	−2.211
	Min.	0.703	0.385	−0.166	2.754	0.374	−2.338
	Max.	0.751	0.412	−0.115	2.889	0.418	−2.097

The load model built from data recorded at substation 6 (Table 2.8), which is one of the 15 distribution substations of Case I used for the test, is polynomial load model. For the accuracy of the obtained load model parameters, taking weekday night load model as an example, the mean square error between the simulated and recorded real power is 5.27×10^{-6} using average parameter values, and it is 5.03×10^{-5}

Table 2.9 Load model parameters of Substation 1, Case I

Parameters		α_t	α_s	T_p	β_t	β_s	T_q
Weekday daytime	Average	2.534	0.643	15.3	4.012	1.233	20.8
	Most probable	2.586	0.632	16.5	4.143	1.216	22.4
	Min.	2.305	0.582	13.5	3.753	0.999	16.7
	Max.	2.778	0.812	17.8	4.287	1.473	24.4
Weekday night	Average	2.467	0.655	16.6	3.745	0.987	24.8
	Most probable	2.425	0.617	16.9	3.784	0.956	25.1
	Min.	2.165	0.536	14.1	3.522	0.748	19.5
	Max.	2.651	0.823	18.4	3.986	1.221	29.2
Weekends	Average	2.356	0.594	15.2	3.464	1.168	18.7
	Most probable	2.388	0.583	14.8	3.422	1.175	18.7
	Min.	2.076	0.437	12.4	3.193	0.974	15.9
	Max.	2.648	0.746	16.5	3.605	1.381	21.3

for reactive power. By using most probable parameter values, the mean square error is 6.73×10^{-6} for real power and 9.64×10^{-5} for reactive power. The simulated responses obtained by load models are plotted in Fig. 2.14 together with recorded responses. Figure 2.15 shows the zoomed plots of real and reactive power around the time of voltage step change.

In Figs. 2.14 and 2.15, black solid line refers to the original measured responses. The black dashed line represents the modified response. The green line with plus signs is the simulated response of this event. The grey area represents the range of all simulated responses whose parameters are obtained from each measured response. Blue dotted and red dash-dot lines are the simulated responses by most probable and average values of all the parameters. As illustrated in the figures, the simulated response can match measured response very well. The simulated response coincides with the modified response.

The load model obtained from Substation 1, (Table 2.9) is exponential dynamic load model. The mean square error between the simulated and measured real power calculated using average parameter values is 9.86×10^{-6}, and it is 6.14×10^{-5} for reactive power. The plots of the recorded response and simulated responses are shown in Fig. 2.16, and zoomed plots in Fig. 2.17. The types of lines representing different responses are the same as in Fig. 2.14. It can be seen that either using average parameters values or the most probable parameter values, the obtained simulated responses match very well the measured response.

The load model derived from Case II (Table 2.10) is composite load model. The mean square error between the simulated and recorded real power is 4.13×10^{-5}, and the value for reactive power is 8.74×10^{-5}. Differently from Case I, where hundreds of events can be used to build one load model, there are only a few events in Case II. Therefore, Fig. 2.18 only shows the comparison of measured and simulated response for one event, Fig. 2.19 shows zoomed section of plots shown in Fig. 2.18. The blue

Table 2.10 Load model parameters, Case II

Parameters	p_1	p_2	p_3	q_1	q_2	q_3	s_0	s_{cr}	t_{m0}	T_j
Value	0.552	0.338	0.110	0.599	−1.427	1.828	1.35	10.29	0.76	0.93

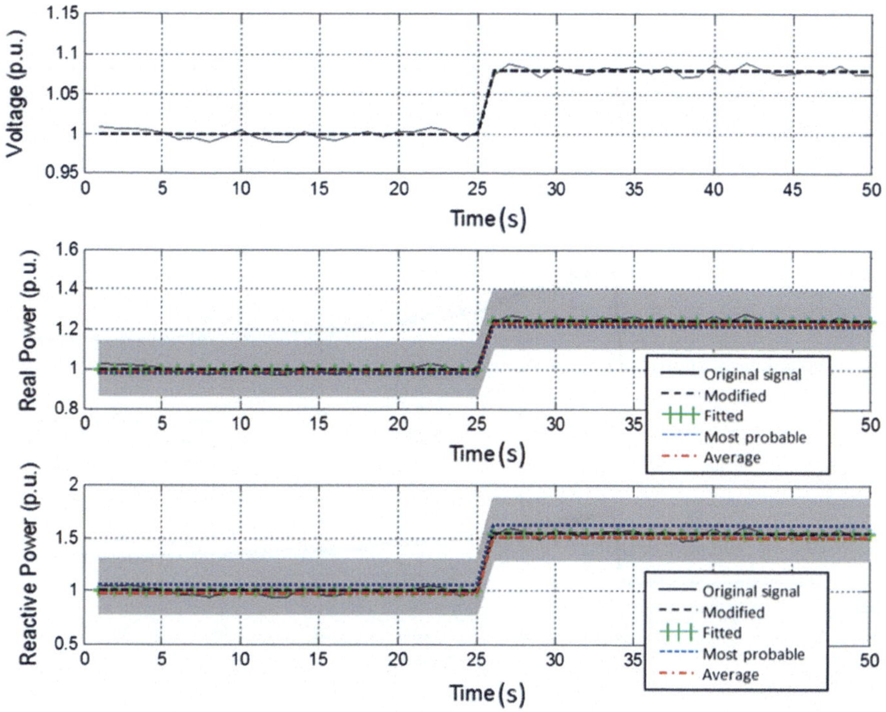

Fig. 2.14 Comparison of simulated response and measured response for polynomial load model

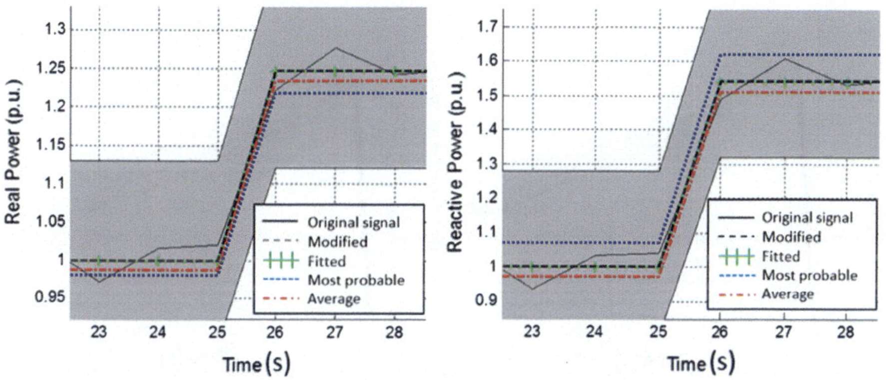

Fig. 2.15 Zoomed plots of real power and reactive power responses around 25 s

solid lines are measured responses, the green dashed lines are modified responses, and the red dotted lines are simulated responses. It can be seen that there is a good match between simulated response and measured response.

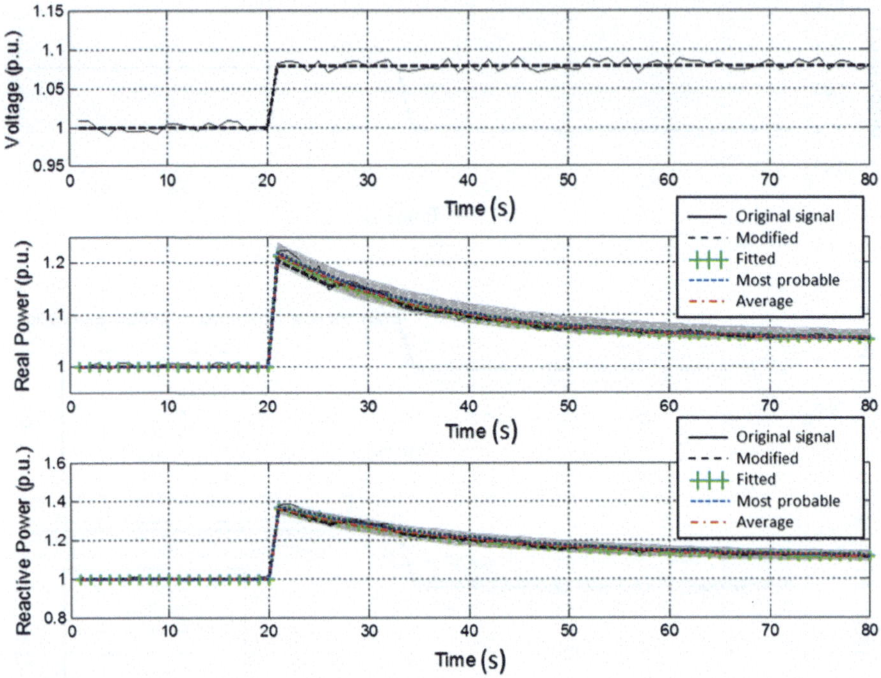

Fig. 2.16 Comparison of simulated response and measured response for exponential dynamic load model

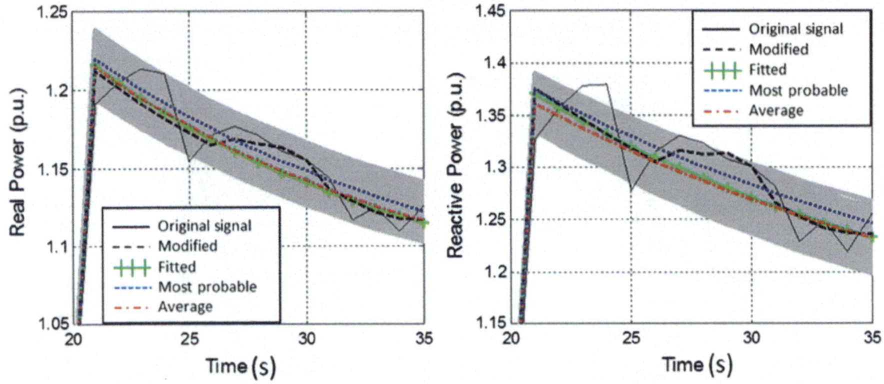

Fig. 2.17 Zoomed plots of real power and reactive power responses between 20 and 35 s

2.3.5 Summary

Reliable power system analysis and control require accurate load models. Measurement based load modelling method is the most commonly used and reliable load

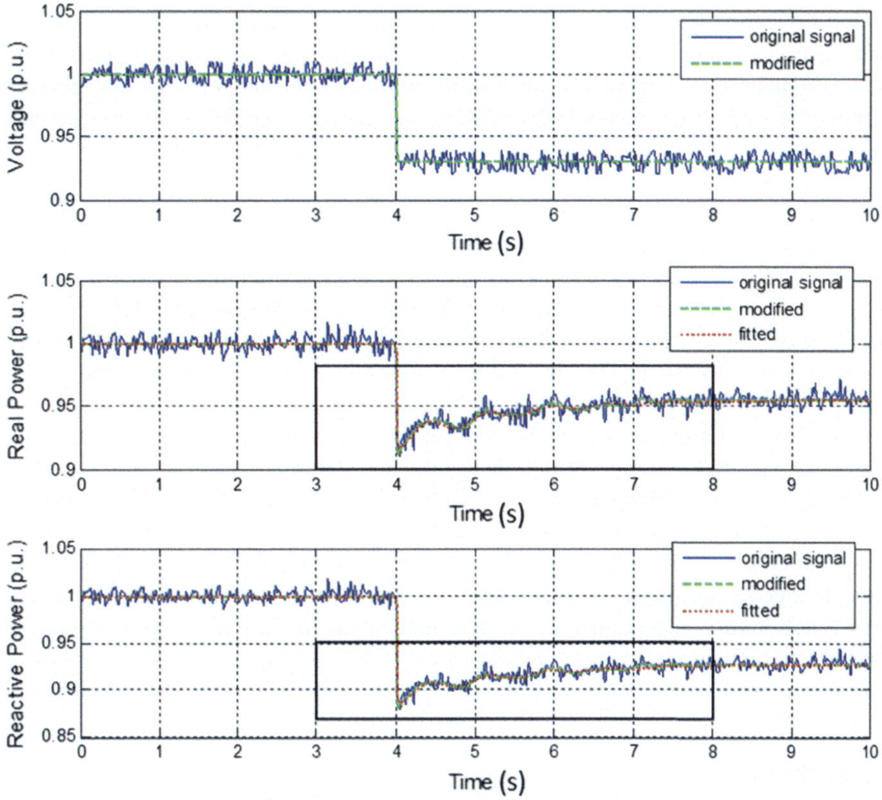

Fig. 2.18 Comparison of simulated response and measured response for composite load model

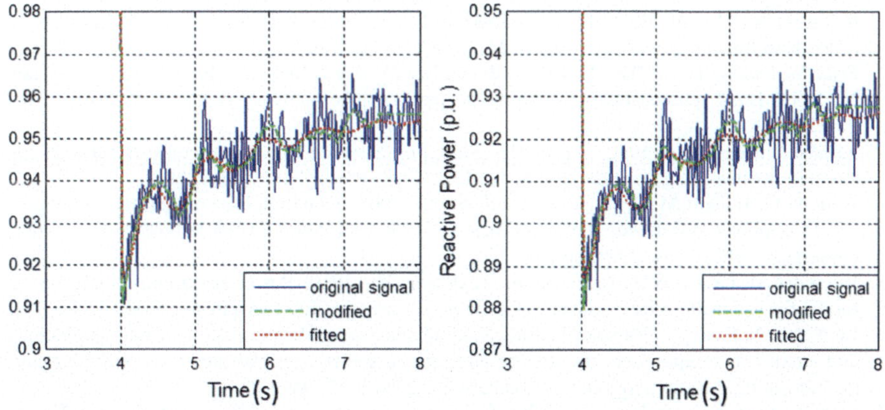

Fig. 2.19 Zoomed plots of real power and reactive power responses between 3 and 8 s

modelling technique. However, this method is time consuming and requires significant resources, and it only accurately represents the load at monitored bus at the time of measurement. Therefore, an automated load modelling tool can significantly improve the efficiency of load modelling, save many resources, and consequently contribute to the reliable assessment of power system stability.

The ALMT developed as a part of this research is fully automated load modelling software, from data input to output of developed load models and corresponding parameters. It represents the first original contribution of this thesis. The ALMT contains three stages. The first stage is data processing, which involves event selection, signal filtering, and event grouping. The second stage is load model selection. In this stage, the most suitable load model is chosen according to the most frequent load response shape. The third stage is parameter fitting, where the load model parameters are fitted by optimisation techniques. The obtained simulated responses are compared with the actual measured load responses and if they don't match, the process is repeated until appropriate load model is found.

The illustrative results presented in the chapter demonstrate that AMLT can automatically develop load models and corresponding parameters with high accuracy from load responses measured in real power systems. This tool can be implemented in existing power quality monitors or fault recorders, which can significantly increase their functionality and contribute to automatic load modelling for power system stability studies.

References

1. Milanovic JV, Gaikwad A, Borghetti A, Djokic SZ, Dong ZY, Andrew Halley, Korunovic LM, Villanueva SN, Ma J, Pourbeik P, Resende F, Sterpu S, Villella F, Yamashita K, Auer O, Karoui K, Kosterev D, Leung SK, Mtolo D, Zali SM, Collin A, Xu Y (2014) Modelling and aggregation of loads in flexible power networks. J. M. CIGRE WG C4.605
2. Price W, Chiang H-D, Clark H, Concordia C, Lee D, Hsu J, Ihara S, King C, Lin C, Mansour Y (1993) Load representation for dynamic performance analysis. IEEE Trans Power Syst 8:472–482
3. Khodabakhchian B, Vuong G-T (1997) Modeling a mixed residential-commercial load for simulations involving large disturbances. IEEE Trans Power Syst 12:791–796
4. Abdalla O, Bahgat M, Serag A, El-Sharkawi M (2008) Dynamic load modelling and aggregation in power system simulation studies. In: 12th international Middle-East power system conference, MEPCON, 2008, pp 270–276
5. Rylander M, Grady WM, Arapostathis A, Powers EJ (2010) Power electronic transient load model for use in stability studies of electric power grids. IEEE Trans Power Syst 25:914–921
6. Lu N, Xie Y, Huang Z, Puyleart F, Yang S, Load component database of household appliances and small office equipment. In: Power and energy society general meeting-conversion and delivery of electrical energy in the 21st century, 2008 IEEE, pp 1–5
7. Gole A, Keri A, Kwankpa C, Gunther E, Dommel H, Hassan I, Marti J, Martinez J, Fehrle K, Tang L (1997) Guidelines for modeling power electronics in electric power engineering applications. IEEE Trans Power Deliv 12:505–514
8. Navarro IR, Samuelsson O, Lindahl S Automatic determination of parameters in dynamic load models from normal operation data. In: Power Engineering Society General Meeting, 2003, IEEE

9. Tomiyama K, Ueoka S, Takano T, Iyoda I, Matsuno K, Temma K, Paserba JJ Modeling of load during and after system faults based on actual field data. In: Power engineering society general meeting, 2003. IEEE, pp 1385–1391
10. Ma J, Han D, He RM, Dong ZY, Hill DJ (2008) Reducing identified parameters of measurement-based composite load model. IEEE Trans Power Syst 23:76–83
11. Choi B-K, Chiang H-D, Li Y, Chen Y-T, Huang D-H, Lauby MG Development of composite load models of power systems using on-line measurement data. In: Power engineering society general meeting, 2006. IEEE, p 8
12. Baghzouz Y, Quist C Composite load model derivation from recorded field data. In: Power engineering society 1999 winter meeting. IEEE, 1999, pp 713–718
13. Keyhani A, Lu W, Heydt GT Composite neural network load models for power system stability analysis. In: Power systems conference and exposition, 2004. IEEE PES, pp 1159–1163
14. Zali SM, Milanović JV (2013) Generic model of active distribution network for large power system stability studies. IEEE Trans Power Syst 28:3126–3133
15. Regulski P, Vilchis-Rodriguez D, Djurović S, Terzija V (2015) Estimation of composite load model parameters using an improved particle swarm optimization method. IEEE Trans Power Deliv 30:553–560
16. Morison K, Hamadani H, Wang L Practical issues in load modeling for voltage stability studies. In: Power engineering society general meeting, 2003, IEEE, pp 1392–1397
17. Louie KW, Marti JR A novel aggregate load model for studying power system dynamic behavior. Presented at the 6th international power engineering conference, Singapore, 2003
18. Smith SW (1997) The scientist and engineer's guide to digital signal processing. California Technical Pub., San Diego, CA
19. Björck Å (1996) Numerical methods for least squares problems: SIAM
20. Irving M, Song Y (2001) Optimisation techniques for electrical power systems-part 2 heuristic optimisation techniques. IEE Power Eng J 15:151–160
21. Sharman K Maximum likelihood parameter estimation by simulated annealing. In: 1988 international conference on acoustics, speech, and signal processing, ICASSP-88, 1988, pp 2741–2744
22. Zhu Y, Milanović JV (2017) Automatic identification of power system load models based on field measurements. IEEE Trans Power Syst
23. Xu W, Vaahedi E, Mansour Y, Tamby J (1997) Voltage stability load parameter determination from field tests on BC Hydro's system. IEEE Trans Power Syst 12:1290–1297
24. Fila M (2010) Modelling, evaluation and demonstration of novel active voltage control schemes to accomodate distributed generation in distribution networks. Brunel University School of Engineering and Design Ph.D. Theses
25. Schafer RW (2011) What is a Savitzky-Golay filter? [lecture notes]. IEEE Signal Process Mag 28:111–117
26. Otto B (2005) Linear algebra with applications. Prentice Hall
27. Mitchell M (1998) An introduction to genetic algorithms. MIT Press
28. Granville V, Krivánek M, Rasson J-P (1994) Simulated annealing: a proof of convergence. IEEE Trans Pattern Anal Mach Intell 16:652–656

Chapter 3
Power System Stability Indices

Building on the results of load model parameters identification presented in Chap. 2, this chapter discusses the methodologies used to assess the influence of load model parameters on power system stability studies. In order to quantitatively measure the effect of parameter variation on the power system stability, stability indices are used to represent the power system stability status. This chapter therefore focuses on description of typical power system stability indices used for this purpose.

3.1 Power System Stability Indices

In order to investigate the effects of load model parameters on different types of power system stability, power system stability indices are needed to quantitatively measure these effects. In the sensitivity analysis, the index value will be the system output. Because for each parameter variation, the index needs to be calculated again, the index should not be too complex to calculate.

3.1.1 Voltage Stability Assessment

The voltage stability limit can be reached when there is a sudden load growth of a relatively small amount during a heavily loaded condition. Therefore an index is needed to indicate the status of voltage stability and the active power increase allowed in order to maintain voltage stability. The PV curve shown in Fig. 3.1 is a practical index for this purpose [1]. The power flow is calculated while increasing the load until the point where power flow solution does not converge [2–4]. The continuation power flow, which can find a continuum of power flow solutions for a given load change scenario, is often used in order to obtain the PV curve [5]. In PV curve, the vertical axis is voltage and the horizontal axis is power. The critical point of the PV curve represents the last point before the voltage stability limit is reached [6]. If the

Y. Zhu, *Power System Loads and Power System Stability*, Springer Theses,
https://doi.org/10.1007/978-3-030-37786-1_3

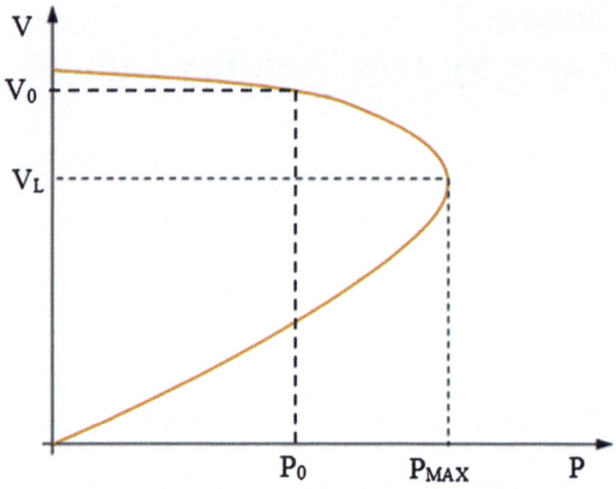

Fig. 3.1 The PV curve

power system is operated at the point P_0 on the upper half of the curve, then the PV margin is the amount of load increase that will cause the power system to reach voltage limit from that particular operating point, as described by Eq. (3.1). In order to find the PV margin, a series of simulations need to be done where the system load will be increased gradually until the load flow equations do not converge. If there is any contingency in the power system, for example, loss of transmission lines, the shape of PV curve will change, and the PV margin might decrease.

$$PV_{margin} = P_{MAX} - P_0 \tag{3.1}$$

3.1.2 *Small Disturbance Stability Assessment*

For small disturbance stability, since the power system equations are linearized and modal analysis is often used to study the small disturbance, the damping of critical electromechanical mode σ_{cr} is used to represent the status of small disturbance stability. A positive σ_{cr} value means that the system is unstable, and a negative σ_{cr} refers to a stable system. If the negative σ_{cr} is closer to 0, then the system stability margin is smaller.

3.1.3 *Transient Stability Assessment*

For transient stability, the Transient Stability Index (TSI), given by Eq. (3.2), is typically used to represent the stability of the system for a specific contingency

[7]. A negative TSI value means that this case is unstable because the rotor angle difference between at least two generators at the same instance is larger than 360 degrees. A positive TSI value represents a stable case. The larger the TSI value, is more stable the system.

$$TSI = 100 \times \frac{360 - \delta_{\max}}{360 + \delta_{\max}} \tag{3.2}$$

where δ_{max} refers to the maximum rotor angle difference between any two generators at the same instance in the system.

3.1.4 Frequency Stability Assessment

For frequency stability, frequency nadir is usually used to represent the status of frequency stability. Frequency nadir is the lowest frequency value after a power system disturbance. Normally, the maximum frequency deviation should not be larger than 2 for 50 Hz system [8, 9].

3.2 Review of Voltage Stability Indices

This section reviews most commonly used indices for the assessment of power system voltage stability. A desirable voltage stability index should be able to measure the amount of load increase that will cause the voltage collapse. In board aspect, these indices can fall into two groups [10]. The first type is based on the power system Jacobian matrix, and capable of calculating the voltage collapse point. The second category applies the admittance matrix and different system variables like load voltages and power angles. The indices in the second group require less calculation time so they are appropriate for online ranking, however they cannot forecast the voltage collapse point. The following subsections give an overview of different indices and their merits and drawbacks.

3.2.1 PV Margins

The PV curve is one of the most commonly used and reliable ways to illustrate the concept of voltage stability. An example of PV curve is illustrated in Fig. 3.1. In practical, The PV curve is obtained by gradually increasing the system loading until the power flow doesn't converge. The shape of PV curve will change, and the PV margin will decrease when there is a contingency in the power system. The PV margin is calculated by (3.1).

The advantages of using PV margin as the voltage stability index are:

- It can directly show the amount of load shedding that needs to be carried out when the pre-fault network load cannot be met even when there is no contingency in the system [6].
- Using the PV curve is the most reliable method to determine the proximity to voltage collapse.

The disadvantage of using loading margin as the voltage stability index is that it is computationally expensive.

3.2.2 L-Indicator

One of the indices for assessing voltage stability which can be implemented in real-time and is conformed to the circuit theory of voltage collapse is called L-indicator [11]. Firstly, it is discussed and explained in two bus system. The simplest form of a system consists of only a generator and a load bus, which is illustrated below (Fig. 3.2).

Focusing on the voltage behaviour of this two bus system.

Let

$$Y_t = \frac{1}{R + jX} \tag{3.3}$$

$$Y_s = K - jM \tag{3.4}$$

$$Y_r = K - jM \tag{3.5}$$

where K is the conductance and M is half of the line susceptance. The subscripts *t, s,* and *r* correspond to transmission, sending end, and receiving end respectively. The sending end represents the generator side, and the receiving end stands for load side.

According to Kirchhoff's current law:

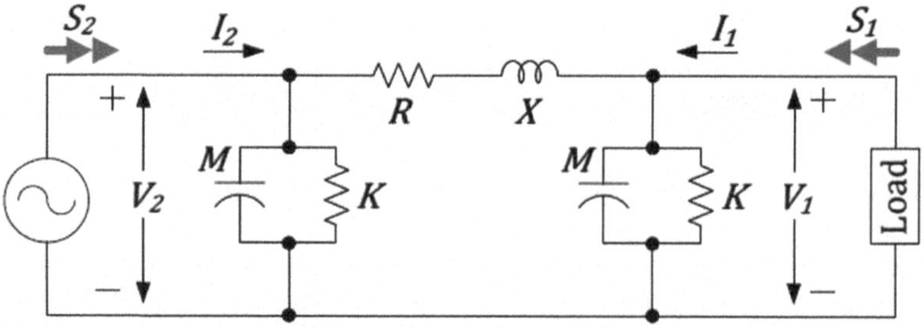

Fig. 3.2 Two-bus system diagram

$$I_1 = I_{Y_r} + I_{Y_t} \tag{3.6}$$

$$I_1 = V_1 Y_r + (V_1 - V_2) Y_t \tag{3.7}$$

and

$$I_1 = \frac{S_1^*}{V_1^*} \tag{3.8}$$

Substituting (3.8) into (3.7) and multiplying by the denominator:

$$S_1^* = V_1^2 Y_r + V_1^2 Y_t - V_1^* V_2 Y_t \tag{3.9}$$

Rewrite (3.9) as:

$$S_1^* = V_1^2 Y_{11} + V_0 V_1^* Y_{11} \tag{3.10}$$

where

$$Y_{11} = Y_r + Y_t \tag{3.11}$$

$$V_0 = \frac{Y_t}{Y_{11}} V_2 \tag{3.12}$$

In order to obtain a solution of load bus voltage, the Eq. (3.10) has to be written in terms of V_1.

Let

$$\frac{S_1^*}{Y_{11}} = a + jb \tag{3.13}$$

Thus, (3.9) becomes

$$a + jb = V_1^2 + V_0 V_1^* \tag{3.14}$$

$$a + jb = V_1^2 + V_0 V_1 \cos(\delta_0 - \delta_1) + j V_0 V_1 \sin(\delta_0 - \delta_1) \tag{3.15}$$

For the real part of (3.15):

$$\cos(\delta_0 - \delta_1) = \frac{a - V_1^2}{V_0 V_1} \tag{3.16}$$

For the imaginary part of (3.15):

$$\sin(\delta_0 - \delta_1) = \frac{b}{V_0 V_1} \tag{3.17}$$

Combining (3.16) and (3.17):

$$V_0^2 V_1^2 = \left(a - V_1^2\right)^2 + b^2 \tag{3.18}$$

Solving (3.18), yields

$$V_1 = \sqrt{\frac{V_0^2}{2} + a \pm \sqrt{\frac{V_0^4}{4} + aV_0^2 - b^2}} \tag{3.19}$$

From (3.13):

$$a = \frac{S_1}{Y_{11}} \cos\left(\theta_{S_1} + \theta_{Y_{11}}\right) \tag{3.20}$$

$$b = -\frac{S_1}{Y_{11}} \sin\left(\theta_{S_1} + \theta_{Y_{11}}\right) \tag{3.21}$$

Rearranging (3.19):

$$V_1 = \sqrt{\frac{V_0^2}{2} + a \pm \sqrt{\left(\frac{V_0^2}{2} + a\right)^2 - \left(a^2 + b^2\right)}} \tag{3.22}$$

Substituting (3.20) and (3.21) into (3.22):

$$V_1 = \sqrt{\frac{S_1}{Y_{11}}\left(r \pm \sqrt{r^2 - 1}\right)} \tag{3.23}$$

where

$$r = \frac{V_0^2 Y_{11}}{2S_1} + cos\left(\theta_{S_1} + \theta_{Y_{11}}\right) \tag{3.24}$$

According to (3.23) the voltage will collapse when $r < 1$. Therefore, when $r = 1$ is the critical state between voltage stable and unstable. When $r = 1$, $\frac{S_1}{V_1^2 Y_{11}} = 1$, and from (3.10) follows:

$$\frac{S_1}{V_1^2 Y_{11}} = 1 + \frac{V_0}{V_1} \tag{3.25}$$

Therefore, the voltage collapse indicator, L, is defined as

$$L \triangleq \left|\frac{S_1}{V_1^2 Y_{11}}\right| = \left|1 + \frac{V_0}{V_1}\right| \tag{3.26}$$

L equals to 0 when there is no load, i.e. $S_1 = 0$. L equals to 1 when voltage collapses. Thus in order to maintain voltage stability, L has to be less than 1.

In order to apply L-indicator to the real power system, it has to be expanded to the multi-bus system. Below is the process of expansion. Firstly, the system is represented by admittance matrix:

$$\begin{bmatrix} I_L \\ I_G \end{bmatrix} = \begin{bmatrix} Y_{LL} & Y_{LG} \\ Y_{GL} & Y_{GG} \end{bmatrix} \begin{bmatrix} V_L \\ V_G \end{bmatrix} \tag{3.27}$$

G and L stand for generator and load buses respectively. Equation (3.27) can be manipulated to the following form.

$$\begin{bmatrix} V_L \\ I_G \end{bmatrix} = \begin{bmatrix} Z_{LL} & F_{LG} \\ K_{GL} & Y_{GG} \end{bmatrix} \begin{bmatrix} I_L \\ V_G \end{bmatrix} \tag{3.28}$$

The jth element of the V_L is

$$V_{Lj} = \sum_{i \in L} Z_{LLji} I_{Li} + \sum_{i \in G} F_{LGji} V_{Gi} \tag{3.29}$$

Multiplying the both side of the equation by V_{Lj}^* yields:

$$\begin{aligned}
V_{L_j}^2 - V_{Lj}^* \sum_{i \in G} F_{LGji} V_{Gi} &= V_{Lj}^* \sum_{i \in L} Z_{LLji} I_{Li} \\
V_{L_j}^2 + V_{0j} V_{Lj}^* &= V_{Lj}^* \left(Z_{LLjj} I_{Lj} + \sum_{\substack{i \in L \\ i \neq j}} Z_{LLji} \frac{S_{L_i}^*}{V_{Li}^*} \right) \\
&= \frac{S_{Lj}^*}{Y_{jj}} + \frac{S_{L_{jCR}}^*}{Y_{jj}} \\
&= \frac{S_{j+}^*}{Y_{jj}}
\end{aligned} \tag{3.30}$$

where

$$S_{L_{jCR}} = \left(\sum_{\substack{i \in L \\ i \neq j}} \frac{Z^*_{LL_{ji}}}{Z^*_{LL_{jj}}} \frac{S_{L_i}}{V_{L_i}} \right) V_{L_i}$$

$$S^*_{j+} = S^*_{Lj} + S^*_{L_{jCR}}$$

The physical understanding of V_{0j} is the contribution of all generators to a specific node, while $S^*_{L_{jCR}}$ represents the contribution of all loads to the same node. Applying the same method as in the two-bus system, the L-indicator for the multi-bus system can be derived as:

$$L_j = \left| 1 + \frac{V_{0_j}}{V_{L_j}} \right| = \left| \frac{S^*_{j+}}{Y_{jj} V^2_{L_j}} \right| \tag{3.31}$$

And the global indicator L used to reflect the voltage stability status of the whole system is defined as:

$$L = MAX\{L_j\} \tag{3.32}$$

The advantages of L-indicator are:

- The local indicator L_j can be used to determine which bus is causing the voltage collapse.
- This index requires less computation time and is therefore more appropriate for fast calculation of power system operating condition.

The disadvantages of L-indicator are:

- The capability of the modelling is limited by the conventional power flow models [12].
- The effect of generators reaching their reactive limits is not considered.

3.2.3 Impedance Ratio Index

The impedance ratio index is a recently proposed index [13]. The following is the process of deriving this index. The figure below is a black box representation of a power system. It has n load buses, m generator buses, and numerous tie-buses within the network (Fig. 3.3).

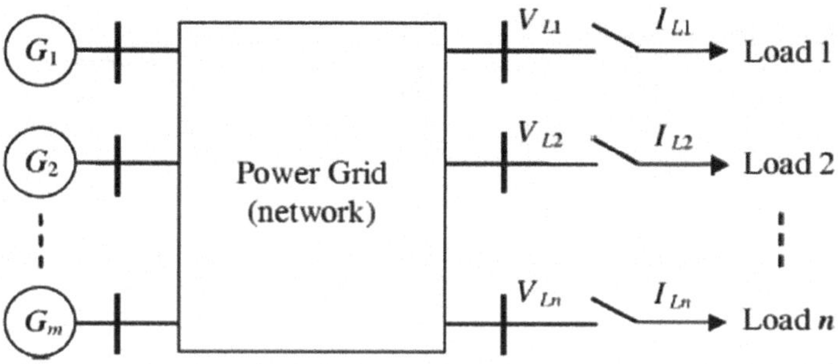

Fig. 3.3 Black box power network representation

Applying nodal analysis, the currents injections for three types of buses are:

$$\begin{bmatrix} -I_L \\ 0 \\ I_G \end{bmatrix} = \begin{bmatrix} Y_{LL} & Y_{LT} & Y_{LG} \\ Y_{TL} & Y_{TT} & Y_{TG} \\ Y_{GL} & Y_{GT} & Y_{GG} \end{bmatrix} \begin{bmatrix} V_L \\ V_T \\ V_G \end{bmatrix} \tag{3.33}$$

where Y is the system admittance matrix, V and I represent the voltage and current vectors, and G, T, and L correspond to buses of generators, ties and loads. According to Eq. (3.33), the load bus voltages can be written as:

$$V_L = KV_G - ZI_L \tag{3.34}$$

where K is a $n \times m$ matrix, and Z is an $n \times n$ impedance matrix. They can be derived from Y matrices. According to Eq. (3.33):

$$V_{L_i} = \sum_{k=1}^{m} K_{ik} V_{G_k} - \sum_{j=1}^{n} Z_{ij} I_{L_j} \quad (i = 1, 2, \ldots, n) \tag{3.35}$$

where K_{ik} denotes the i–k element of K, V_{G_k} represents the voltage of the kth generator, Z_{ij} represents the i–j element of the impedance matrix Z, V_{L_i} and I_{L_j} stand for the voltage and current of load i, respectively. Observing Eq. (3.34), it can be found out that $\sum_{k=1}^{m} K_{ik} V_{G_k}$ stands for the open-circuit voltage of load i. Then the above equation can be rewritten as:

$$V_{L_i} = E_{open,i} - Z_{eq,i} I_{L_j} \quad (i = 1, 2, \ldots, n) \tag{3.36}$$

where

$$E_{open,i} = \sum_{k=1}^{m} K_{ik} V_{G_k} \quad (i = 1, 2, \ldots, n) \tag{3.37}$$

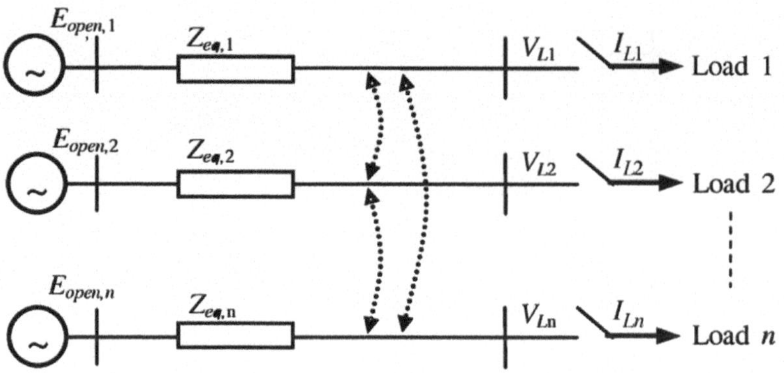

Fig. 3.4 The decoupled equivalent system

$$Z_{eq,i} = \sum_{j=1}^{n} Z_{ij} \frac{I_{L_j}}{I_{L_i}} \quad (i = 1, 2, \ldots, n) \tag{3.38}$$

Equation (3.36) represents a two-bus equivalent circuit seen from load *i*, and $Z_{eq,i}$ and $E_{open,i}$ stand for the equivalent impedance and voltage respectively. Thus the power network can be decoupled into *n* single equivalent circuits seen from individual load buses, as illustrated in Fig. 3.4.

It can be proven that the load impedance Z_{L_i} matches the equivalent impedance $Z_{eq,i}$ at the maximum system loading [14].

$$Z_{Li} = \sum_{j=1}^{n} Z_{ij}^{*} \frac{I_{L_j}}{I_{L_i}} \quad (i = 1, 2, \ldots, n) \tag{3.39}$$

The impedance ratio voltage stability index is defined as:

$$r_i = \left| \frac{1}{Z_{Li}} \sum_{j=1}^{n} Z_{ij}^{*} \frac{I_{L_j}}{I_{L_i}} \right| \quad (i = 1, 2, \ldots, n) \tag{3.40}$$

Since Z_{Li} cannot be measured directly, (3.40) is rewritten as:

$$r_i = \left| \frac{1}{V_{Li}} \sum_{j=1}^{n} Z_{ij}^{*} I_{L_j} \right| \quad (i = 1, 2, \ldots, n) \tag{3.41}$$

The numerator of (3.41) represents the sum of voltage drops resulted from the load currents in the system. r_i can act as an indicator of the voltage stability of load *i*. As the loading increases, the value of r_i will be higher. Thus, the load *i* is more voltage stable with a smaller r_i value.

The advantages of Impedance ratio index is that it can be applied to fast online indication of voltage stability. The disadvantage is that it assumes loads to be constant impedance load, which is not always the case.

3.2.4 Voltage Collapse Index

This index is proposed after the investigation of the way of directly estimating the bifurcation value λ^* of a saddle-node bifurcation [15]. The proposed voltage collapse index is the estimation of the margin from the present value λ_1 to λ^*. To derive this index, first consider a power system described by

$$\dot{\boldsymbol{x}} = f(\boldsymbol{x}, \lambda) \tag{3.42}$$

where $\boldsymbol{x}$ is a state vector including bus voltage magnitudes and angles, and λ is a parameter vector including real and reactive power at each load bus. If the parameter λ changes slowly or quasi-statically with respect to the dynamics of Eq. (3.42), then the above equation can be approximated by:

$$0 = f(\boldsymbol{x}, \lambda) \tag{3.43}$$

Let

$$\boldsymbol{J}_{lk} \triangleq \left(\boldsymbol{I} - \boldsymbol{e}_1\boldsymbol{e}_1^T\right)\boldsymbol{J} + \boldsymbol{e}_1\boldsymbol{e}_k^T \tag{3.44}$$

where $\boldsymbol{J} = \frac{\partial f}{\partial x}$, $\boldsymbol{I}$ represents the identity matrix, $\boldsymbol{e}_i$ stands for the *ith* unit vector, and indices l and k are chosen to ensure $\boldsymbol{J}_{lk}$ is non-singular. Next, consider the following system:

$$\boldsymbol{J}_{lk}(\boldsymbol{x}, \lambda)\boldsymbol{v} = \boldsymbol{e}_l \tag{3.45}$$

where $\boldsymbol{e}_l$ is the lth unit vector. Let $\bar{\boldsymbol{v}}$ be the solution vector of the above equation and $(\boldsymbol{x}, \lambda) = (\boldsymbol{x}_1, \lambda_1)$. Equation (3.45) can be obtained from the value of the following test function:

$$\begin{cases} \boldsymbol{J}(\boldsymbol{x}, \lambda)\boldsymbol{v} = 0 \\ \boldsymbol{e}_k^T \boldsymbol{v} = 1 \end{cases} \tag{3.46}$$

Considering the following test function:

$$t_s(\boldsymbol{x}, \lambda) \triangleq \boldsymbol{e}_1^T \boldsymbol{J}(\boldsymbol{x}, \lambda)\boldsymbol{v} \tag{3.47}$$

Equation (3.47) is used as the test function because it is a function of $\boldsymbol{x}$ and λ and at the bifurcation point λ^* it is zero. The derivative of (3.47) is:

$$t_s' = \boldsymbol{e}_1^T \left(\boldsymbol{J}' \boldsymbol{v} + \boldsymbol{J} \boldsymbol{v}' \right) \tag{3.48}$$

Now the question is obtaining the value λ^* such that the value of the test function $t_s(\boldsymbol{x}^*, \lambda^*)$ is zero when the operating point $(\boldsymbol{x}_1, \lambda_1)$ is provided. Simulation results show that when a change of real or reactive demand on a single bus happens, the test function can be approximately quadratic. Also, when a change of both real and reactive demands on multiple buses occurs, the test function can be best fit by a quartic model. These properties lead to the following expression for the λ^*.

$$\lambda^* \approx \lambda_1 - \frac{1}{c} \frac{t_s(\boldsymbol{x}_1, \lambda_1)}{t_s'(\boldsymbol{x}_1, \lambda_1)} \tag{3.49}$$

where $c = 2$ or 4 for the quadratic or quartic models respectively. The voltage collapse index is then defined as follows:

$$\tau = \lambda^* - \lambda_1 \tag{3.50}$$

This index can be used as an approximation of the exact load margin for voltage stability.

The advantages of the voltage collapse index are:

- There is a direct relationship between the value of the index and the amount of load change that will lead to instability.
- To calculate the voltage collapse index, it only involves calculating two load flow solutions and a few matrix computations. Thus it can be applied to fast online application.

The disadvantages of the voltage collapse index are:

- This index does not consider some physical constraints like OLTCs and VAR limits.
- The crucial bus can only be found when the system nearly reach the bifurcation point [16].

3.2.5 *Channel Components Transform (CCT)*

The power system can be seen as a multimode, multi-branch Thevenin circuit that connects the loads with generators [17]. The Thevenin equivalent circuit of a general power network can be described in the following form:

$$V = KE - ZI = E' - ZI \tag{3.51}$$

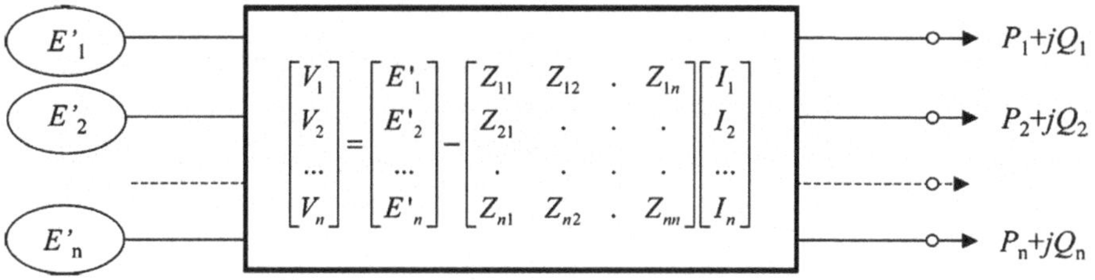

Fig. 3.5 A multimode, multibranch Thevenin circuit, adopted from [17]

where E is the vector of terminal voltages of the generators and V is the vector of nodal voltages at load buses. $E' = KE$ is the open circuit voltage vector of the traditional multimode Thevenin equivalent circuit. Figure 3.5 is a general circuit model for the representation of Eq. (3.51).

Applying eigenvalue decomposition on Z yields:

$$Z = T^{-1}\Lambda T \tag{3.52}$$

where T and Λ are the eigenvector and eigenvalue matrices of Z. Substituting (3.52) in (3.51) yields:

$$\begin{aligned} V &= KE - ZI = KE - T^{-1}\Lambda TI \\ TV &= TKE - \Lambda TI \end{aligned} \tag{3.53}$$

$U = TV$ is the transformed voltage

$J = TI$ is the transformed current

$F = TKE$ is the transformed voltage source

Then the network can be expressed by the following decoupled form:

$$\begin{bmatrix} U_1 \\ U_2 \\ \cdots \\ U_n \end{bmatrix} = \begin{bmatrix} F_1 \\ F_2 \\ \cdots \\ F_n \end{bmatrix} - \begin{bmatrix} \lambda_1 & 0 & 0 & 0 \\ 0 & \lambda_2 & 0 & 0 \\ 0 & 0 & \cdots & 0 \\ 0 & 0 & 0 & \lambda_N \end{bmatrix} \begin{bmatrix} J_1 \\ J_2 \\ \cdots \\ J_n \end{bmatrix} \tag{3.54}$$

This representation is illustrated in Fig. 3.6:

The advantage of CCT lies in that it can decouple the supply system into independent channel circuits. However, when it is applied to physically decoupled loads, the channel loads become coupled. The reason is that CCT is a linear transform, while a load is a nonlinear variable. Assuming the loads are constant impedance loads, then the physical load at different buses can be given as:

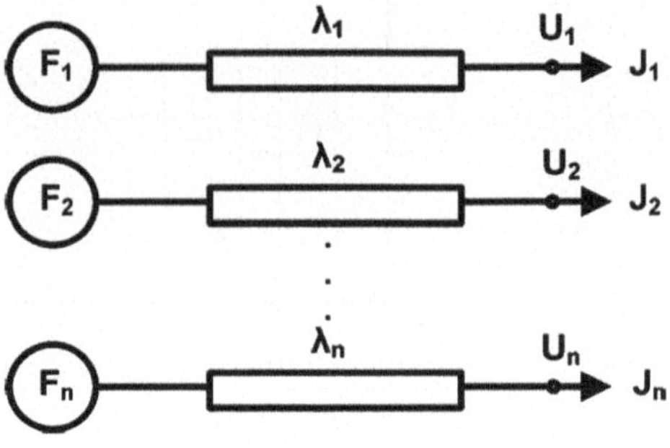

Fig. 3.6 Decoupled representation of a complex network, adopted from [17]

$$\begin{bmatrix} V_1 \\ V_2 \\ \cdots \\ V_n \end{bmatrix} = \begin{bmatrix} Z_{L1} & 0 & \cdots & 0 \\ 0 & Z_{L2} & & 0 \\ & \vdots & \ddots & \vdots \\ 0 & 0 & \cdots & Z_{Ln} \end{bmatrix} \begin{bmatrix} I_1 \\ I_2 \\ \cdots \\ I_n \end{bmatrix} \tag{3.55}$$

In the channel domain, the loads become

$$V = Z_L I$$
$$U = T Z_L T^{-1} J = Z_C J$$

$$\begin{bmatrix} U_1 \\ U_2 \\ \cdots \\ U_n \end{bmatrix} = \begin{bmatrix} Z_{C11} & Z_{C12} & \cdots & Z_{C1n} \\ Z_{C21} & Z_{C22} & & Z_{C2n} \\ \vdots & & \ddots & \vdots \\ Z_{Cn1} & Z_{Cn2} & \cdots & Z_{Cnn} \end{bmatrix} \begin{bmatrix} J_1 \\ J_2 \\ \cdots \\ J_n \end{bmatrix} \tag{3.56}$$

Equation (3.56) shows that loads are coupled in the channel domain. In order to decouple the channel loads, a change has to be made to the channel voltage source F_i:

$$F_{eqi} = F_i - \lambda_i J_{Ei} \tag{3.57}$$

All channels are decoupled from each other after that, which makes it easy to examine the PV curves from the channel domain. For this purpose, the relationship between the channel voltage and the channel load power as shown in Fig. 3.7 is derived as:

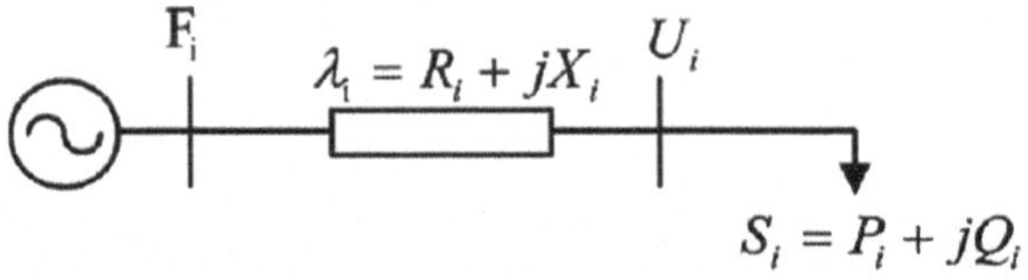

Fig. 3.7 The ith channel circuit

$$U_i^4 + U_i^2\left[2(P_iR_i + Q_iX_i) - F_i^2\right] + |\lambda_i|^2\left(P_i^2 + Q_i^2\right) = 0 \tag{3.58}$$

A critical channel exists among all different channels of a system which is responsible for the voltage collapse. The channel that has the smallest channel margin is identified as the critical channel. To obtain the critical channel, the maximum channel power is given as:

$$S_{max} = \frac{\left|F_{eqi}\right|^2[|\lambda_i| - (X_i\sin\theta_i + R_i\cos\theta_i)]}{2[X_i\cos\theta_i - R_i\sin\theta_i]^2} \tag{3.59}$$

where θ_i is the power factor angle of the ith channel load. The channel margin can then be calculated by:

$$Channel\, margin = \frac{S_{\max} - S_{operating}}{S_{operating}} \tag{3.60}$$

where $S_{operating}$ is the channel load power at the current operating point.

After the critical channel is obtained, it can be used to identify the critical bus. The critical bus is the bus that caused the largest voltage drop of the critical channel because the low stability margin of the critical channel corresponds to its large channel voltage drop. Since the channel current is responsible for the voltage drop, the contribution of bus currents I to the critical channel current J_i can be used as the index for determining critical bus. The current J_i of the critical channel is given as:

$$Ji = T_{i1}I_1 + T_{i2}I_2 + \ldots + T_{iN}I_N \tag{3.61}$$

The index for load ranking is then:

$$Cont_{ik} = \frac{|T_{ik}I_k|\cos(\alpha_{ik})}{|J_i|} \tag{3.62}$$

where $Cont_{ik}$ represents the contribution of load bus k to channel current J_i and α_{ik} denotes the angle between the J_i and $T_{ik}I_k$. The load bus with the largest value of $Cont_{ik}$ is regarded as the critical load bus.

The advantages of the Channel Components Transform Index are as follows:

- This index is relatively easy to calculate since it does not involve calculating power flow.
- There is clear physical meaning and models for channel components and circuits, which makes it easy to extract useful information and interpret the results in channel domain.

The disadvantage of the Channel Components Transform Index is that the index assumes the loads to be constant impedance load model, which will limit its application.

3.2.6 Diagonal Element Dependent Index

The value of diagonal elements $\frac{\partial Q_i}{\partial V_i}$ and $\frac{\partial P_i}{\partial \delta_i}$ of Jacobian Matrix decline when there is an increase in load at bus i [18]. Therefore, the change of value $\frac{\partial Q_i}{\partial V_i}$ and $\frac{\partial P_i}{\partial \delta_i}$ from the no-load value to a specific loading condition value could be applied as the voltage stability index for the load bus i. The detailed derivation is given in [18]. I_{pi} is defined as the value of $\frac{\partial P_i}{\partial \delta_i}$ at the specified loading condition over the value of $\frac{\partial P_i}{\partial \delta_i}$ at no-load condition. I_{qi} is defined as the value of $\frac{\partial Q_i}{\partial V_i}$ at the specified loading condition over the value of $\frac{\partial Q_i}{\partial V_i}$ at no-load condition. Both denominators can be replaced by $-B_{ii}$, where B is the imaginary part of the admittance matrix Y.

$$I_{qi} = \frac{\partial Q_i/\partial V_i}{-B_{ii}} \tag{3.63}$$

$$I_{pi} = \frac{\partial P_i/\partial \delta_i}{-B_{ii}} \tag{3.64}$$

However, those two indices are not very reliable in indicating the voltage instability point because they do not have a constant threshold value. Therefore, it leads to another index:

$$I_i = \frac{I_{pi}}{V_{Geq}} \tag{3.65}$$

where

$$V_{Geq} = \sum_{\substack{j=1 \\ j \neq i}}^{N} \frac{B_{ij}}{-B_{ii}} V_j \tag{3.66}$$

$$I_i = \frac{\partial P_i/\partial \delta_i}{\sum_{j=1, j\neq i}^{N} B_{ii} V_j} \tag{3.67}$$

The advantage of the Diagonal element Dependant Index is that I_i not only reduce with increase in load, but also its threshold value is close to 0.5 in most situations, thus it is more reliable in indicating the distance to voltage instability. The disadvantage of the Diagonal element Dependant Index is that the load that is closest to the voltage collapse will be regarded as the most critical bus, which normally should be the load that has the largest influence on the voltage stability.

3.2.7 *Comparison of Different Voltage Stability Indices*

Although five different voltage stability indices have been discussed in this Section, they have not been adapted for the parameter ranking for voltage stability. The main reason is that the voltage stability analysis was needed to be performed together with other stability studies in the same standardised simulation environment, i.e., in DIgSILENT PowerFactory. In PowerFactory however there is not enough information available to implement some of these indices, most notably, there is no access to system Jacobian Matrix. Another reason is that classical PV margin itself, is a very reliable voltage stability index that can be calculated easily and fast in PowerFactory. Therefore, the PV margin is used for assessment of the influence of the load model parameters on voltage stability.

A comparison of six voltage stability indices is performed in MATLAB R2013a on the modified version of a reduced order equivalent model of NETS-NYPS test system (New England Test System—New York Power System) [19]. The diagram of the system is shown in Fig. 3.8.

Figure 3.9 illustrates the value of six voltage stability indices as the loading increases from normal loading condition. The voltage collapse point is reached when the system loading is 1.45 p.u. In order to fit different indices into the same figure, the values of indices are normalised between 0 and 1.

Figure 3.9 shows that the L-indicator (L), Impedance Ratio Index (Ri), and Diagonal element dependent index (I) have no advantage in indicating the distance to voltage collapse compared with PV curve. Their values all changes rapidly in

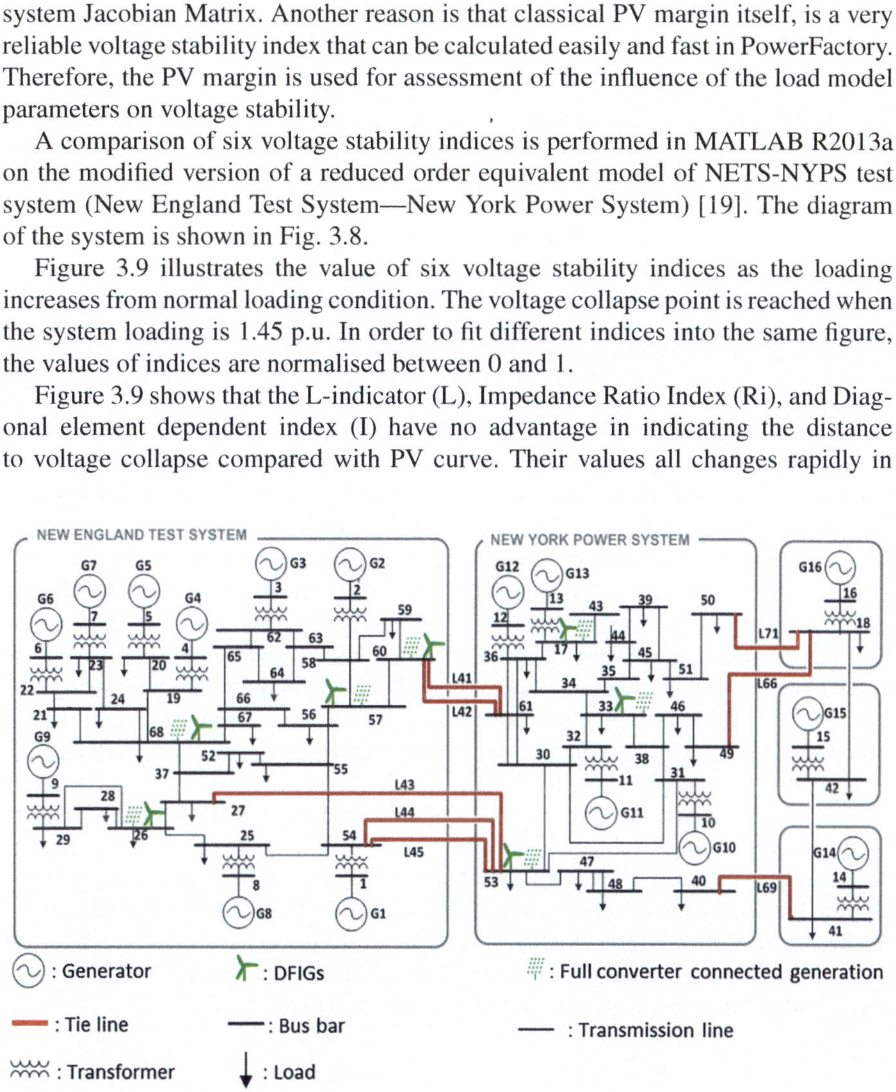

Fig. 3.8 The NETS-NYPS test system

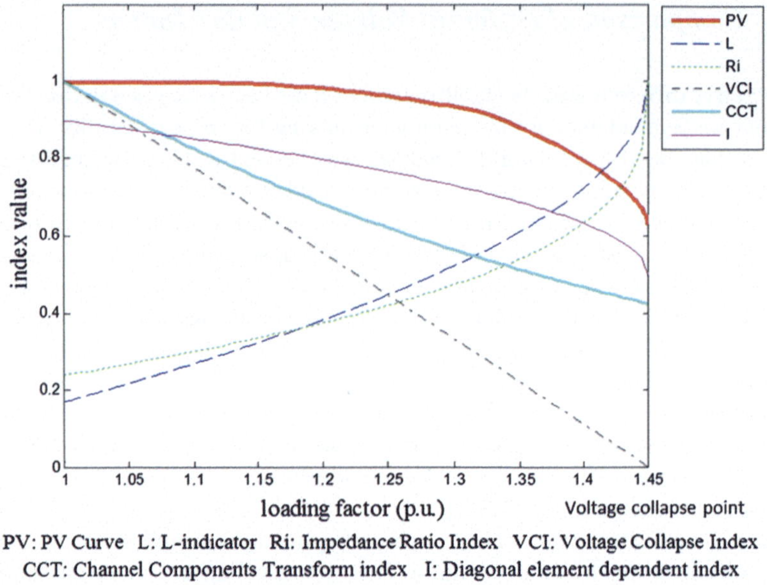

Fig. 3.9 Comparison of different voltage stability indices

the vicinity of voltage collapse point. The I index has slight benefit (more reliable in indicating the distance to voltage instability) of the instability point being near I = 0.5, but considering the strong nonlinearity near the instability point, this information is not that useful. Although the calculation time required to assess these indices compared to PV curve may be quicker, they cannot be realised in DIgSILENT PowerFactory environment. The Channel Components Transform index (CCT) looks smoother than previous four indices, but it is not directly used for load model parameter ranking. The contribution of the load bus to the channel current still need to be calculated, which will be much more complex compared with directly using PV curve. Another disadvantage is that one cannot know when the voltage collapse point will be reached by only looking at channel current value. Also, CCT is only limited to constant impedance load model. Finally, the VCI index shows a very good linear property, as it is actually an approximation of the load margin. This index seems to be a very effective voltage stability index. However, the limitation is that loading factor must increase very slowly, which may not always be true, and the calculation of VCI requires Jacobian Matrix, which cannot be accessed in DIgSILENT PowerFactory. The Jacobian Matrix is not directly accessible to users because there are a lot of internal algorithms in place to optimise the power flow calculation. This is also the case for other commercial power system analysis software.

3.3 Review of Small Disturbance Stability Indices

This section reviews most commonly used indices for the assessment of power system small disturbance stability. A desirable small disturbance stability index should be able to indicate how close the power system operating condition is to small disturbance stability limit. The following subsections give an overview of different indices. They are both based on the linear analysis of power system equations.

3.3.1 Damping of Critical Mode

Recalling from Sect. 1.2.3, for small disturbance stability, the power system equations can be linearised as (1.4) and (1.5). When there are non-trivial solutions to (3.68), the scalar λ represents the eigenvalues of $\boldsymbol{A}$, where $\boldsymbol{\phi}$ is an $n \times 1$ vector and $\boldsymbol{\phi} \neq 0$. There are n solutions to (3.71), corresponding to n eigenvalues $\lambda = \lambda_1, \lambda_2 \ldots \lambda_n$.

$$\boldsymbol{A\phi} = \lambda\boldsymbol{\phi} \tag{3.68}$$

$e^{\lambda_i t}$ represents the time-based behaviour of a mode λ_i [20]. Therefore if the eigenvalue does not have imaginary part, then it is non-oscillatory. If the real part is negative, then it will have a decaying time response. If the real part is positive, then it will have an increasing time response. If eigenvalues are complex, they will only appear in conjugate pairs $\lambda = \sigma \pm j\omega$, where σ is damping and ω is frequency. If any mode in a system has a positive real part, then the system will be unstable. Therefore, the damping of the mode that is closest to the imaginary axis in complex plane (critical mode) is used to represent the distance to instability.

3.3.2 Damping Factor of Critical Mode

In some situations, the damping factor ζ is used instead of damping as the small disturbance stability index. The damping factor of a mode is given by (3.69):

$$\zeta = \frac{-\sigma}{\sqrt{\sigma^2 + \omega^2}} \tag{3.69}$$

It describes the rate of decay of the oscillation amplitudes associated with the mode. If the damping factor has a positive value, then these oscillations will decay and tend to steady state. If it has a negative value, then these oscillations will increase and lead to system instability. Therefore, a large positive damping factor value is desired, often with a threshold of $\zeta > 5\%$ implemented for designing of power system control [21].

This index is basically equivalent to damping of critical mode. It is more useful though when the speed of damping of oscillation is the focus of the study. The damping of the critical mode however more directly indicates the status of small disturbance stability, thus it is used in this study.

3.4 Review of Transient Stability Indices

This section reviews five indices for the assessment of power system transient stability. They are divided into two groups. The first one includes the transient stability index (TSI) and transient rotor angle severity index (TRASI), which reflect the stability status of the whole system. The second one comprises of the generator specific indices, i.e., maximum rotor angle deviation, maximum speed deviation and maximum acceleration.

3.4.1 Transient Stability Index

The transient stability index (TSI) is a frequently used index for representing the stability of the system after a contingency [7]. The expression for calculating TSI is given by (3.2). A negative TSI value means that the system is unstable and a positive TSI value means that the system is stable. In addition, a larger TSI value represents a more stable system.

3.4.2 Transient Rotor Angle Severity Index

The transient rotor angle severity index (TRASI) is defined in (3.70) [22]:

$$TRASI = \frac{360^{\circ} - \delta_{\max}^{pst}}{360^{\circ} - \delta_{\max}^{pre}} \tag{3.70}$$

where $\delta_{\max}^{pre}$ and $\delta_{\max}^{pst}$ are the pre-disturbance maximum rotor angle difference and post-disturbance maximum rotor angle difference respectively.

TRASI measures the severity of rotor angle separation after a fault in the system by comparing the maximum rotor angle difference before and after the disturbance. The value of the index ranges between 0 and 1. A larger index indicates more stable system as it means that the rotor angle difference after the disturbance is smaller.

3.4.3 Generator Specific Indices

The TSI compares the rotor angle value of all generators in the system, while the following three indices focus on individual generators. They are maximum rotor angle deviation, maximum speed deviation and maximum acceleration. The expressions for these indices are given by (3.71–3.73), respectively.

$$\Delta\delta_i^{\max} = \max\left(\left|\delta_i - \delta_i^0\right|\right) \tag{3.71}$$

$$\Delta\omega_i^{\max} = \max(|\Delta\omega_i(t)|) = \max\left(\left|\frac{\delta_i(t) - \delta_i(t - \Delta t)}{2\pi f \Delta t}\right|\right) \tag{3.72}$$

$$\Delta a_i^{\max} = \max(|\Delta a_i(t)|) = \max\left(\left|\frac{\omega_i(t) - \omega_i(t - \Delta t)}{\Delta t}\right|\right) \tag{3.73}$$

where δ_i^0 represents the initial rotor angle of each generator, ω_i stands for rotor speed, and a_i is the acceleration of rotor speed. For these indices, it is required to specify which generator is looked into.

3.4.4 Comparison of Different Transient Stability Indices

By comparing three generator specific indices with TSI and TRASI, it can be found that although the generator specific indices can provide more information about the characteristics of each generator, for example, some generators have a higher value of speed deviation, these indices are not as useful as TSI and TRASI in indicating the stability status of the system. The main reason is that they do not have a specific threshold value for instability. Whether the system is stable or not, or how close the system is to instability, cannot be read from generator specific indices.

After comparing TSI and TRASI, it can be seen that they are both determined by the maximum rotor angle difference after the disturbance ($\delta_{\max}^{pre}$ can be seen as a constant). If represented in p.u., their values are both between 0 and 1 for stable operating conditions, and negative for unstable operating conditions. Using both indices leads to the same conclusion about system stability. However, TSI has been more commonly used in the past research about transient stability [7, 23–26]. Therefore, the practice of using TSI as the transient stability index is followed in this thesis.

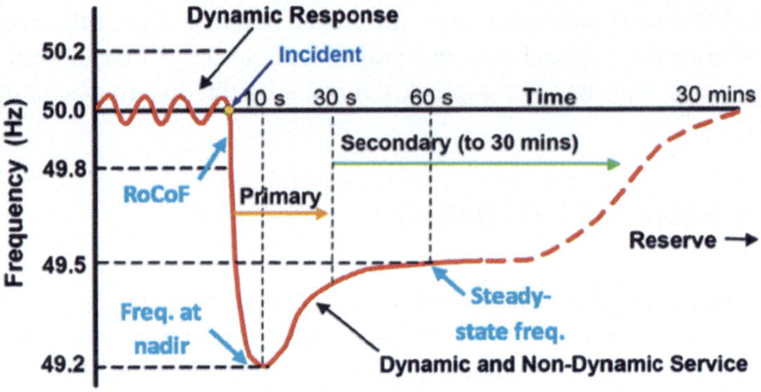

Fig. 3.10 Power system frequency response after a disturbance, adopted from [34]

3.5 Review of Frequency Stability Indices

In this section, two frequency stability indices have been reviewed—frequency nadir and rate of change of frequency (ROCOF). After a brief comparison, the frequency nadir is chosen to be the index used in this thesis.

3.5.1 Frequency Nadir

The most commonly used frequency stability index is frequency nadir [22, 27–30]. It is the lowest frequency value after a power system disturbance, as shown in Fig. 3.10. The system is more stable if the frequency nadir is closer to normal operating frequency i.e., 50 or 60 Hz. Normally, the maximum frequency deviation should not be larger than 2 for 50 Hz system [8, 9]. This index can be easily obtained from system measurement devices.

3.5.2 Rate of Change of Frequency

The rate of change of frequency (ROCOF) can also reflect the frequency stability of a power system. From the swing equation, the ROCOF can be expressed as [31]:

$$\frac{\partial f}{\partial t} = \frac{\Delta P}{2H} \tag{3.74}$$

where ΔP is the active power disturbance, and H is system inertia. The ROCOF is also illustrated in Fig. 3.10. A greater ROCOF indicates that the system is less stable, or rather more prone to instability.

Although in many cases, both indices are used to analyse the frequency response more comprehensively [28, 31], it is beneficial to focus on one index in order to quantitatively measure the frequency stability. The frequency nadir is chosen in this study to assess system frequency stability for two reasons. First reason is its wide use in studies about frequency stability [8, 22, 27, 32, 33], showing that it is a very reliable index. The second reason is that ROCOF only reflects the initial frequency response after the disturbance, and that after that the response will be heavily affected by different types of controls. The frequency nadir on the other hand represents the lowest frequency level attained considering the whole response. Therefore, the frequency nadir is a more suitable index to quantitatively measure the frequency stability.

3.6 Summary

In order to quantitatively measure power system stability, this chapter reviewed the power system stability indices used for different stability studies in the past research. A comparison of these indices has been made, and the most suitable index has been chosen for each type of stability. The chosen indices are PV margin for voltage stability, damping of critical mode for small disturbance stability, TSI for transient stability, and frequency nadir for frequency stability.

References

1. Suzuki M, Wada S, Sato M, Asano T, Kudo Y (1992) Newly developed voltage security monitoring system. Power Syst IEEE Trans 7:965–973
2. Capitanescu F, Van Cutsem T (2002) Preventive control of voltage security margins: a multicontingency sensitivity-based approach. Power Systems, IEEE Transactions on 17:358–364
3. Nallan H, Rastgoufard P (1996) Computational voltage stability assessment of large-scale power systems. Electr Power Syst Res 38:177–181
4. Braz L, Castro C, Murati C (2000) A critical evaluation of step size optimization based load flow methods. Power Syst IEEE Trans 15:202–207
5. Ajjarapu V, Christy C (1991) The continuation power flow: a tool for steady state voltage stability analysis. In: [Proceedings] conference papers 1991 power industry computer application conference. pp 304–311
6. Greene S, Dobson I, Alvarado FL (1997) Sensitivity of the loading margin to voltage collapse with respect to arbitrary parameters. Power Syst IEEE Trans 12:262–272
7. Papadopoulos PN, Milanović JV (2017) Probabilistic framework for transient stability assessment of power systems with high penetration of renewable generation. IEEE Trans Power Syst 32:3078–3088
8. Seethalekshmi K, Singh SN, Srivastava SC (2011) A synchrophasor assisted frequency and voltage stability based load shedding scheme for self-healing of power system. IEEE Trans Smart Grid 2:221–230
9. Boemer JC, Burges K, Zolotarev P, Lehner J, Wajant P, Fürst M, Brohm R, Kumm T (2011) Overview of German grid issues and retrofit of photovoltaic power plants in Germany for the prevention of frequency stability problems in abnormal system conditions of the ENTSO-E

region continental Europe. In: 1st international workshop on integration of solar power into power systems
10. Karbalaei F, Soleymani H, Afsharnia S (2010) A comparison of voltage collapse proximity indicators. In: IPEC, 2010 Conference Proceedings. pp 429–432
11. Kessel P, Glavitsch H (1986) Estimating the voltage stability of a power system. Power Eng Rev IEEE PER-6:72–72
12. Gao B, Morison G, Kundur P (1992) Voltage stability evaluation using modal analysis. Power Syst IEEE Trans 7:1529–1542
13. Xiaoming M, Weixing L, Tiankui S, Shuang R, Yong S (2013) An analytical sensitivity index for load shedding to avoid voltage instability. Power Energy Soc General Meet (PES) IEEE 2013:1–5
14. Weixing L, Tongwen C (2009) An investigation on the relationship between impedance matching and maximum power transfer. Electr Power Energy Conf (EPEC) IEEE 2009:1–6
15. Hsiao-Dong C, Jean-Jumeau R (1995) Toward a practical performance index for predicting voltage collapse in electric power systems. Power Syst IEEE Trans 10:584–592
16. Cañizares C, De Souz AC, Quintana VH (1996) Comparison of performance indices for detection of proximity to voltage collapse. Power Syst IEEE Trans 11:1441–1450
17. Xu W, Pordanjani IR, Wang Y, Vaahedi E (2012) A network decoupling transform for phasor data based voltage stability analysis and monitoring. Smart Grid IEEE Trans 3:261–270
18. Sinha A, Hazarika D (2000) A comparative study of voltage stability indices in a power system. Int J Electr Power Energy Syst 22:589–596
19. Preece R (2013) Improving the stability of meshed power networks: a probabilistic approach using embedded HVDC lines. Springer Science & Business Media
20. J. Machowski, J. Bialek, J. R. Bumby, and J. Bumby, *Power system dynamics and stability*: John Wiley & Sons, 1997
21. Rogers G (2012) Power system oscillations. Springer Science & Business Media
22. Meegahapola L, Flynn D (2010) Impact on transient and frequency stability for a power system at very high wind penetration. Power Energy Soc General Meet IEEE 2010:1–8
23. Morales JD, Papadopoulos PN, Milanović JV (2017) Feasibility of different corrective control options for the improvement of transient stability. PowerTech IEEE Manchester 2017:1–6
24. Gautam D, Vittal V, Harbour T (2009) Impact of increased penetration of DFIG-based wind turbine generators on transient and small signal stability of power systems. IEEE Trans Power Syst 24:1426–1434
25. Gomez FR, Rajapakse AD, Annakkage UD, Fernando IT (2011) Support vector machine-based algorithm for post-fault transient stability status prediction using synchronized measurements. IEEE Trans Power Syst 26:1474–1483
26. Mochamad RF, Preece R (2017) Impact of model complexity on mixed AC/DC transient stability analysis
27. O'Sullivan J, Rogers A, Flynn D, Smith P, Mullane A, O'Malley M (2014) Studying the maximum instantaneous non-synchronous generation in an island system—frequency stability challenges in Ireland. IEEE Trans Power Syst 29:2943–2951
28. Wu L, Infield D (2014) Power system frequency management challenges–a new approach to assessing the potential of wind capacity to aid system frequency stability. IET Renew Power Gener 8:733–739
29. Mu Y, Wu J, Ekanayake J, Jenkins N, Jia H (2013) Primary frequency response from electric vehicles in the Great Britain power system. IEEE Trans Smart Grid 4:1142–1150
30. Yan R, Saha TK, Modi N, Masood N-A, Mosadeghy M (2015) The combined effects of high penetration of wind and PV on power system frequency response. Appl Energy 145:320–330
31. Adrees A, Milanovic JV (2016) Study of frequency response in power system with renewable generation and energy storage. Power Syst Comput Conf (PSCC) 2016:1–7

32. Horne J, Flynn D, Littler T (2004) Frequency stability issues for islanded power systems. Power Syst Conf Exposition IEEE PES 2004:299–306
33. Tang J, Liu J, Ponci F, Monti A (2013) Adaptive load shedding based on combined frequency and voltage stability assessment using synchrophasor measurements. IEEE Trans Power Syst 28:2035–2047
34. Teng F, Mu Y, Jia H, Wu J, Zeng P, Strbac G (2016) Challenges of primary frequency control and benefits of primary frequency response support from electric vehicles

Chapter 4
Probabilistic Assessment and Sensitivity Analysis in Stability Studies

Power system parameters, operating conditions, and disturbances are stochastic. When using deterministic simulation, only one operating condition is considered, and the uncertainties neglected. The deterministic study often considers the worst operating condition, thus cannot accurately assess the power system stability. Besides, an increasing number of stochastic renewable generation sources and new types of loads have been connected to power system, introducing more uncertainties and imposing more challenges to stable operation of the power system.

4.1 Probabilistic Modelling of Power System Uncertainties

There are two types of uncertainties that need to be modelled for reliable system stability studies. The uncertainties in system components include the availability of components, and the parameter uncertainty including load demand, power generated by renewable generation, parameters of components, etc. [1]. In addition, power system uncertainties can be divided into two types: random and non-random. The random uncertainties have known probability distribution, such as load demand, generation costs, and power generated by renewables which can be assessed based on historic data. The non-random uncertainties like uncertainty in generation expansion have statistics that cannot be derived from the past observations [2].

Power system variables can be modelled by different types of stochastic distributions. Table 4.1 lists some of the most commonly used probabilistic distributions of different power system variables in literature. The corresponding mathematical formulations of pdf for the chosen probability distributions, and the meaning of parameters of these probabilistic distributions are also included in Table 4.1.

Y. Zhu, *Power System Loads and Power System Stability*, Springer Theses,
https://doi.org/10.1007/978-3-030-37786-1_4

Table 4.1 Probabilistic distributions of power system uncertainties

Uncertainties	Probabilistic Distributions	Equation	Parameters
Power system load	Normal [3–7]	$f(x) = \frac{1}{\sigma\sqrt{2\pi}} e^{-\frac{(x-\mu)^2}{2\sigma^2}}$	μ: mean σ: standard deviation
Wind power	Weibull [8–11]	$\frac{\alpha}{\beta^\alpha} x^{\alpha-1} e^{-\left(\frac{x}{\beta}\right)^\alpha}$	α: scale parameter β: shape parameter
Solar power	Beta [3, 4]	$\frac{(x-a)^{p-1}(b-x)^{q-1}}{B(p,q)(b-a)^{p+q-1}}$	a: first shape parameter b: second shape parameter
Fault occurrence	Poisson [12]	$\frac{e^{-\lambda}\lambda^x}{x!}$	λ: mean
Fault location	Uniform [12]	$\frac{1}{n}$	n: maximum observable value
Fault clearing time	Normal [13]	$f(x) = \frac{1}{\sigma\sqrt{2\pi}} e^{-\frac{(x-\mu)^2}{2\sigma^2}}$	μ: mean σ: standard deviation

4.2 Probabilistic Simulation Method

Using probabilistic simulation methods enables the researcher to analyse the power system considering the uncertainties. The system behaviour can be better understood and predicted by simulating stochastic behaviour of power system parameters. The Monte Carlo (MC) method is the most common and most frequently used probabilistic simulation method. The Monte Carlo method provides an appropriate probabilistic procedure for investigating complex large-scale power systems with high dimensional space of input random variables. The theoretic base for applying Monte Carlo method in uncertainty analysis are two fundamental theorems of statistics: The Weak Law of large numbers [14], which states that the sample average converges in probability towards the expected value and The Central Limit Theorem [15], which states that when independent random variables are added, their properly normalised sum tends toward a normal distribution.

In Monte Carlo simulations, a large amount of random sampling of system uncertainties is performed. From these sampling, the distribution of the power system output variables can be obtained. To perform a Monte Carlo simulation, one has to determine the range of possible inputs, generate inputs randomly according to a certain probabilistic distribution function, calculate the outputs for each sampling, and combine all the results. This is a highly flexible method and does not have limits. It has been applied in many previous studies [16–19].

4.2.1 Simulation Requirements for Monte Carlo Simulation

As stated above, Monte Carlo method is a sampling procedure based on pdf. Random samples of size N are generated according to the pdfs of the input variables. In order to save calculation time, a stopping rule is needed to determine the minimum number of simulations required to achieve some specified confidence level. The sequential estimation is needed to determine this minimum number [20]. The following are the steps for obtaining this minimum simulation requirement [15]:

Suppose the sampling is made according to following procedure:

1. Take n_O samples and record $X_1, \ldots, X_{n0}$; $n_0 \rightarrow N$.
2. Continue sampling until $\frac{(T_N+1)}{N} \leq \frac{\varepsilon^2 N}{\beta_N^2}$, with $N+1 \rightarrow N$, and record X_N for *Nth* sample. Then calculate:

$$\overline{X}_N = \frac{1}{N}\sum_{i=1}^{N} X_i \tag{4.1}$$

$$T_N = \sum_{i=1}^{N}(X_i - \overline{X}_N)^2 \tag{4.2}$$

3. Stop sampling, and then:

$$\lim_{\varepsilon_r \to 0} \frac{N}{n_0(|\mu|\varepsilon_r, \delta)} = 1 \tag{4.3}$$

$$\lim_{\varepsilon_r \to 0} pr\left(\left|\overline{X}_N - \mu\right| \leq |\mu|\varepsilon_r\right) = 1 - \delta \tag{4.4}$$

$$\lim_{\varepsilon_r \to 0}\left(\frac{E(N)}{n_0(|\mu|\varepsilon_r, \delta)}\right) = 1 \tag{4.5}$$

$$\beta = \Phi^{-1}\left(1 - \frac{\delta}{2}\right) \tag{4.6}$$

where ε_r is the relative error for the results, β is the maximum uncertainty level that should be achieved, n_0 is the initial iteration value, N is the actual sample size needed to achieve the desired accuracy ε_r with the coverage probability 1–δ as ε_r tends to 0. $\overline{X}$ is the mean value of the result, μ is the real mean value of the studies variables, and Φ^{-1} is the inverse function of the normal distribution with $\sigma = 1$ and $\mu = 0$. Equation (4.3) shows that the number of iterations N must fulfil the condition that the relative error (ε_r, δ) converges with probability one to the deterministic repetition amount $n_0(|\mu|\varepsilon_r, \delta)$. Also, Eq. (4.4) shows that the desired accuracy ε_r is asymptotically achieved by the sequential procedure with the desired coverage probability $1 - \delta$ as ε_r tends to 0, namely asymptotic consistency. An expression for the relative error can be derived based on the procedure above as [21]:

$$\varepsilon_r = \frac{\Phi^{-1}(1 - \delta/2) \times \sqrt{\sigma'_N/N}}{E(N)} \tag{4.7}$$

where σ'_N and $E(N)$ are the variance and mean of the obtained results respectively. In Monte Carlo simulation, the relative error ε_r is calculated in each simulation and compared with a target relative error. The simulation stops when ε_r is less than the target relative error. It has to be noted that this formula has its limitation. It can be seen from the derivation that Eq. (4.7) can only be strictly applied to samples that follow normal distribution.

4.3 Sensitivity Analysis Techniques

4.3.1 Comparison of Different Sensitivity Analysis Techniques

There are three types of sensitivity analysis techniques: (i) local, (ii) screening, and (iii) global. The computation time increases from (i) to (iii).

The local sensitivity techniques investigate how an individual input parameter affects the model output locally. The computation time for this method is short. It only needs n + 1 model evaluations for an n uncertainties system. However, this technique is not accurate when the system is nonlinear and the uncertainties are large [22, 23].

The screening method can identify the influential uncertain parameters, but cannot provide the information of their contribution to the output. This method requires much less calculation time than the global method, and has much better performance than the local sensitivity analysis techniques [22–24].

In comparison, the global method ranks the uncertain parameters considering all possible input values. It is complex to implement and requires a much longer calculation time [22, 23].

After the comparison of these three methods, it can be concluded that if the system is very complex, like a power system with thousands of buses, it would be very time consuming to use the global method. However, the local method is not reliable enough. Under such circumstances, the screening method is the best choice. A comprehensive comparison of these three types of sensitivity analysis techniques for application in power system small disturbance stability studies has been made in [25], and the advantages of screening method were clearly shown. One of such method is Morris Screening Method. The Morris method showed very good performance in the assessment of wastewater treatment, and global warming [26, 27].

4.3.2 *Morris Screening Method*

The Morris screening method can identify important uncertain variables with a simulation time much shorter than the global method. The uncertain parameter whose changes will cause the largest variation in output will be regarded as the most influential parameter. By using this principle, the Morris method can rank the uncertain parameters and also their impact on the power system stabilities. The procedure for implementing the Morris screening method is given as follows:

Only one variable is varied at one time by a magnitude of Δ. The elementary effect of the ith factor at a given point x is

$$d_i(\boldsymbol{x}) = \frac{[y(x_1, \ldots, x_{i-1}, x_i + \Delta, x_{i+1}, \ldots, x_k) - y(\boldsymbol{x})]}{\Delta} \quad (4.8)$$

where $\mathbf{x}$ is a point in the input region selected such that the perturbed point $\mathbf{x} + \Delta$ is still in input region. d_i is the elementary effect, and Δ is the step size. The size of the step Δ is determined as a multiple of 1/(r–1), where r is the number of variations for each variable. The results will be more accurate for a larger value of r, but r is normally not exceeding 10 [25].

A trajectory is constructed for each Morris simulation. In each step, only one variable is varied by Δ, and the value of other variables remain the same. Following this, another variable is selected randomly, and another step of Δ is made. The whole process finishes when each variable has r steps. Therefore, an input trajectory in the uncertain parameter space is obtained. This is shown as an example in Fig. 4.1.

The starting point is randomly selected for all uncertain parameters between Δ and 1–Δ. The direction of each step is selected randomly as long as the next step

Fig. 4.1 An example trajectory in the input factor space for k = 3 and r = 3

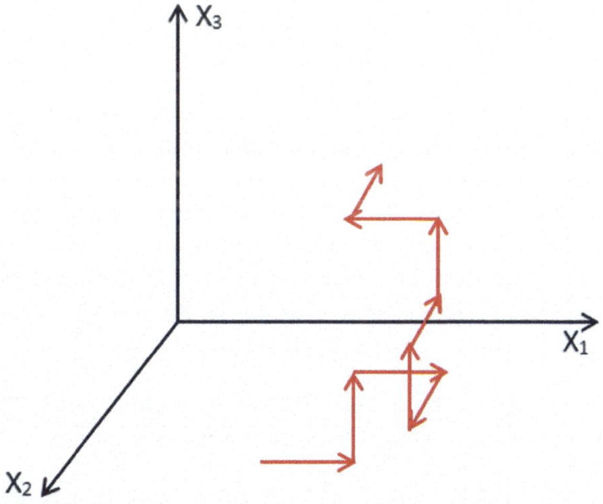

does not exceed the value range of each parameter. The red line in Fig. 4.1 is one possible trajectory for three uncertain parameters and ten steps.

There are two indices that can be used to judge the importance of uncertain parameters. One is the mean μ, another is the standard deviation σ of the elementary effects for each input variable.

$$\mu = \frac{1}{r}\sum_{i=1}^{r}|d_i| \tag{4.9}$$

$$\sigma = \sqrt{\frac{1}{r}\sum_{i=1}^{r}(d_i - \mu)^2} \tag{4.10}$$

where μ represents the sensitivity of the output with respect to ith input variable. If μ is higher, it means that uncertain parameter is more influential. σ represents the standard deviation of the value of the elementary effects. If an uncertain parameter has a large σ, then it means that its influence on the output is nonlinear, and it has correlation with other variables. One of the major advantages of the Morris method is that compared with global sensitivity technique, it requires much less computation time. For k parameters and r levels, it only needs kr + 1 simulations.

4.3.3 *Application Example of Morris Screening Method*

This subsection provides a simple example of how to apply the Morris screening method. A simple mathematical model is used, and detailed procedure is described. The model has 3 input parameters x_1, x_2, and x_3. The function to be evaluated is:

$$y = x_1^2 + 5x_2 - 2x_3 \quad (0 \le x_1, x_2, x_3 \le 8) \tag{4.11}$$

Let the base value for (x_1, x_2, x_3) be (4,4,4), and the level r = 5. The step size Δ is calculated by multiplying the value of the range of inputs (8 in this case) by 1/(r–1), that is $\Delta = 8 \times \frac{1}{5-1} = 2$. Therefore, $3 \times 5 + 1 = 16$ simulations are needed for this simple case study. Table 4.2 records the input and output values of each step change.

In order to find the sensitivity of x_1, it is needed to identify first the steps in which x_1 has changed, i.e., step 3, 6, 8, 11, 13, and then to record the output of that step and the previous step, for example, 30 and 42 for step 3. The mean value of the elementary effects of x_1 is then:

$$\mu = \frac{|30-42| + |36-24| + |46-26| + |60-32| + |32-56|}{5 \times 2} = 9.6 \tag{4.12}$$

where 30 is the output of step 3 and 42 is the output of step 2, 36 is the output of step 6 and 24 is the output of step 5 and so on. Similarly, the mean of elementary effects

Table 4.2 The elementary effects of each step change

Step	Input values	Output values	Step	Input values	Output values
0	(4,4,4)	28	8	(6,2,0)	46
1	(4,6,4)	38	9	(6,2,2)	42
2	(4,6,2)	42	10	(6,0,2)	32
3	(2,6,2)	30	11	(8,0,2)	60
4	(2,6,0)	34	12	(8,0,4)	56
5	(2,4,0)	24	13	(6,0,4)	32
6	(4,4,0)	36	14	(6,2,4)	42
7	(4,2,0)	26	15	(6,2,6)	38

of x_2 and x_3 are 5 and 2 respectively. Therefore, x_1 is the most influential parameter and x_3 is the least influential parameter, which is as expected considering the form of the example function (4.11).

4.4 Stochastic Dependence of Uncertain Parameters

In real power systems, load model parameters are actually correlated with each other, and these correlations are nonlinear and non-Gaussian [28]. It is therefore necessary to consider the mutual dependence of load model parameters as independent probability distributions and random sample generation techniques do not take correlation into account, which may results in significant errors [29–34].

4.4.1 Pearson Correlation Coefficient

The Pearson correlation coefficient (PCC) measures the linear correlation between two variables X and Y [35]. The value of PCC is between 1 and -1, where 1 is total positive linear correlation, 0 represents no linear correlation, and -1 stands for total negative linear correlation. The following figure shows the different scatter plots with different PCC values (Fig. 4.2).

The PCC is calculated by dividing the covariance of two variables by the product of their standard deviations. For two data sets $x_1, \ldots, x_n$ and $y_1, \ldots, y_n$ containing n samples, the PCC is calculated by (4.13):

$$PCC = \frac{\sum_{i=1}^{n}(x_i - \bar{x})(y_i - \bar{y})}{\sqrt{\sum_{i=1}^{n}(x_i - \bar{x})^2}\sqrt{\sum_{i=1}^{n}(y_i - \bar{y})^2}} \tag{4.13}$$

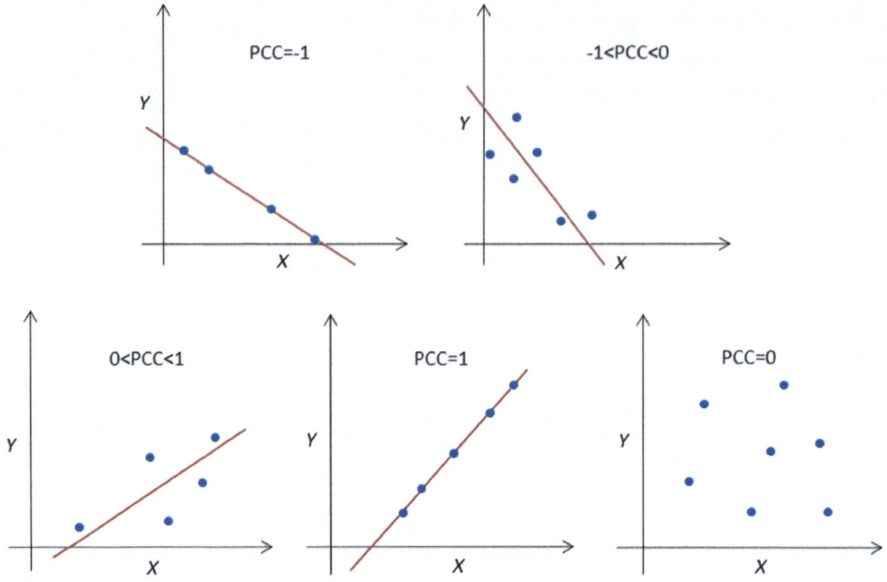

Fig. 4.2 Scatter plots with different PCC values

where n is the size of the sample, x_i and y_i are individual sample points, and $\bar{x}$ and $\bar{y}$ are sample means. There exist several guidelines for the interpretation of a correlation coefficient [36, 37]. However, the interpretation of a correlation coefficient depends heavily on the variables been studied.

4.4.2 Kernel Density Estimation

The kernel density estimation (KDE) is used to estimate the pdf of a random variable in a non-parametric way [38]. Based on a finite data sample, KDE infers the population in the way of data smoothing. Let $x_1, x_2, \ldots, x_n$ be a sample of a distribution with an unknown pdf f. The estimation of f made by KDE is given by Eq. (4.14):

$$\hat{f}_h(x) = \frac{1}{nh}\sum_{i=1}^{n} K\left(\frac{x - x_i}{h}\right) \tag{4.14}$$

where K is the kernel, a non-negative function whose integral is one. h is a positive smoothing parameter called the bandwidth. The commonly used kernel functions are: normal, Epanechnikov, triweight, biweight, triangular, and uniform [39]. In terms of mean square error, the Epanechnikov kernel has the optimal performance [40]. However, normal kernel has convenient mathematical properties and the loss of efficiency is small compared with Epanechnikov kernel [39], thus normal kernel is more frequently used. For normal kernel:

$$K(x) = \phi(x) \tag{4.15}$$

where ϕ represents the standard normal density function. Kernel density estimates are closely related to histograms, but the results are more smooth and continuous. To illustrate the difference, the construction of histogram and kernel density estimators are compared in the following example using six data points (−2.1, −1.3, −0.4, 1.9, 5.1, 6.2).

For the histogram, the horizontal axis is divided into bins covering the range of the data. For this sample, it is divided into 6 bins with a width of 2. Each data point is represented by a box of height $\frac{1}{6\times2} = \frac{1}{12}$. When more than one data point falls into a same bin, the boxes are stacked on top of each other.

For the kernel density estimation, a normal kernel with variance of 2.25 (given by MATLAB function ks density using example data points) is placed on each of the data points x_i, represented by red dashed lines. The kernel density estimation presented by solid blue curve is made by summing the kernels. The histogram and kernel density estimation are illustrated in Fig. 4.3. The smoothness of the kernel density estimate is evident compared to the discreteness of the histogram, as kernel density estimates converge faster to the true underlying density for continuous random variables.

The bandwidth h has a significant effect on the resulting estimation. To illustrate the influence of bandwidth, s simulated random sample from the standard normal distribution is taken, plotted as blue spikes on the horizontal axis in Fig. 4.4. The grey curve represents the true density, which is a standard normal distribution. The red curve is the estimation using a bandwidth of 0.05. It is under-smoothed because it has too many spurious peaks, thus the bandwidth of 0.05 is too small. The green curve is the estimation obtained with a bandwidth of 2. It is over-smoothed because it is too even compared with the true pdf, thus the bandwidth of 2 is too large. The

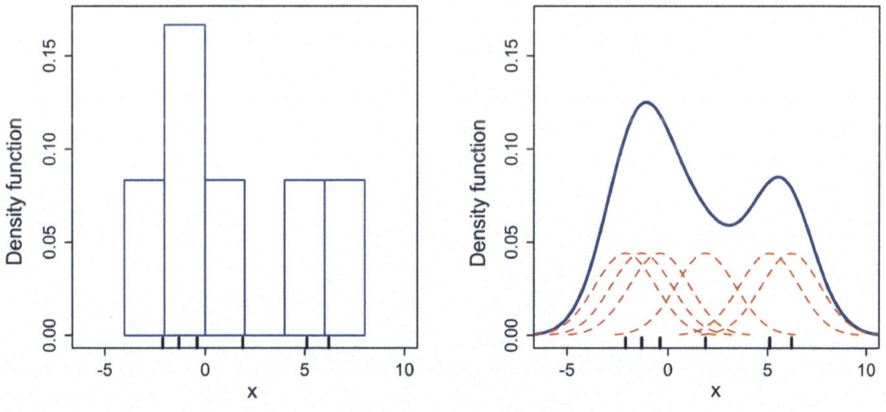

Fig. 4.3 Comparison of histogram and kernel density estimation

Fig. 4.4 KDE obtained using different bandwidths

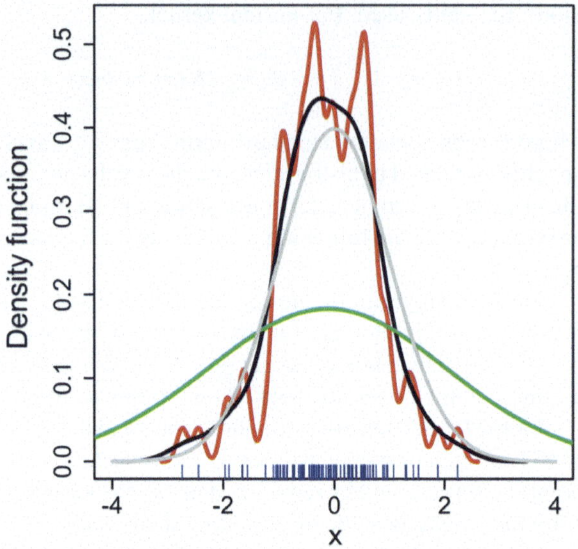

black curve is the estimation with a bandwidth of 0.337. It is very close to the true pdf, thus 0.337 is considered as the optimal bandwidth.

The mean integrated squared error (MISE) is the most frequently used criterion to select the optimal bandwidth, which is given by:

$$MISE(h) = E\left[\int \left(\hat{f}_h(x) - f(x)\right)^2 dx\right] \tag{4.16}$$

where f is unknown real pdf. Under weak assumptions on f and K, MISE can be expressed by:

$$MISE(h) = AMISE(h) + o\left(\frac{1}{nh} + h^4\right) \tag{4.17}$$

where o is the little o notation, defined as:

$$\lim_{h \to \infty} \frac{o\left(\frac{1}{nh} + h^4\right)}{\frac{1}{nh} + h^4} = 0 \tag{4.18}$$

and AMISE is the Asymptotic MISE given by:

$$AMISE(h) = \frac{R(K)}{nh} + \frac{1}{4} m_2(K)^2 h^4 R(f'') \tag{4.19}$$

where

$$R(g) = \int g(x)^2 dx \tag{4.20}$$

$$m_2(K) = \int x^2 K(x) dx \tag{4.21}$$

The minimum of AMISE is the solution to the differential equation:

$$\frac{\partial}{\partial h} AMISE(h) = -\frac{R(K)}{nh^2} + m_2(K)^2 h^3 R(f'') = 0 \tag{4.22}$$

which leads to:

$$h_{AMISE} = \frac{R(K)^{1/5}}{m_2(K)^{2/5} R(f'')^{1/5} n^{1/5}} \tag{4.23}$$

However, both AMISE and h_{AMISE} cannot be used directly because the unknown pdf f or its second derivative f'' are involved. Therefore, many automatic, data-based methods have been developed for finding the optimal bandwidth [41–44].

4.4.3 Importance of Stochastic Dependence of Load Model Parameters

In power systems, load model parameters are actually correlated with each other, and these correlations are nonlinear and non-Gaussian [28]. It is necessary to consider the dependence of load model parameters because independent probability distributions and random sample generation techniques do not take correlation into consideration, which may results in significant errors [29–34]. The stochastic dependence among parameters cannot be truly reflected by linear correlation, which may results in non-optimum solutions of power system studies. In comparison, Copula theory is one effective way of modelling stochastic dependence among parameters.

Different copula families are used to reveal different dependence structures among correlated parameters. Among these copula families, the most frequently used are Archimedean and elliptical copula [45]. Archimedean copula, including Clayton, Frank, and Gumbel copula, is appropriate for reflecting complex dependence structure, but only for two dimensions. For higher dimensions, elliptical copula (i.e., Gaussian, Student's t) should be used, which can represent multiple marginal distributions.

The structure of dependence among uncertain variables, such as symmetry, asymmetry, tail dependence etc., decides which copula should be used. The correlation among wind, photovoltaics, and load have been previously modelled by bivariate methods, including Gumbel, Clayton, and Frank copulas [34, 46, 47]. Multivariate copulas like Gaussian and Student's t copulas have also been applied to power systems studies in [29, 34, 46, 48]. Some other methods, such as joint distributions, pair copula, vine copula, and dependent discrete convolution have also been applied for representing stochastic dependence [30, 31, 33, 49]. But these methods are not widely used because they are complex and computational demanding.

The stochastic dependence modelling have been used in load LVRT (low-voltage ride-through) applications [47], demand response [34], transmission planning [29], generation scheduling [48], and load flow [46]. The applications of copula theory in power systems are limited because it is complex to implement. The application of copula theory in power system stability assessment is especially rare. No previous work has modelled the stochastic dependence of load model parameters using copula theory to assess their impacts on power system stability.

4.4.4 Modelling Stochastic Dependence

In order to model the stochastic dependence of load model parameters, the static exponential load model parameters k_p and k_q of 20 load buses have been obtained from actual, yearlong field measurements. The Pearson correlation coefficients between k_p and k_q for 20 buses are illustrated in a 40×40 matrix presented in Fig. 4.5. In Fig. 4.5, row/column 1–20 stand for k_p of bus 1–20, and row/column 21–40 stand for k_q of substation 1–20. It can be seen that high correlation coefficient values exist between the k_p and k_q of the same bus.

The copula theory is then applied to model the correlation among these parameters. In copula modelling, there are four consecutive stages for generating correlated

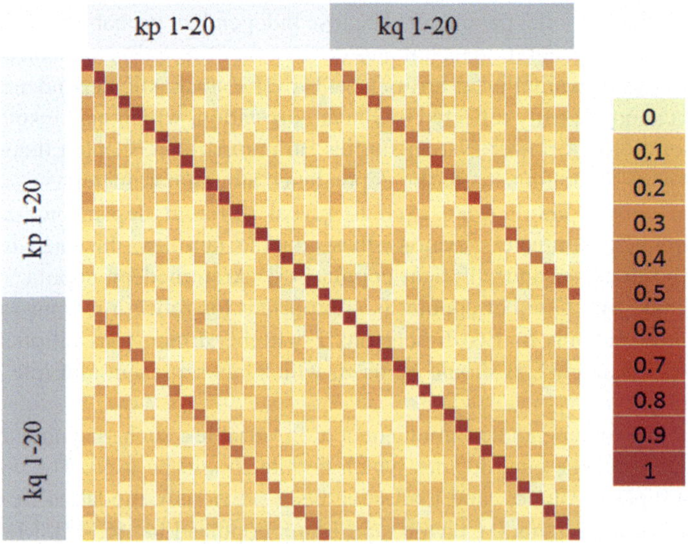

Fig. 4.5 Correlation coefficient matrix of load model parameters

samples from raw data sets: (i) the data is transformed to the unit square by a kernel estimator of the cumulative distribution function, (ii) fitting a copula to the data to obtain the copula parameter, (iii) using copula to generate correlated sample, and (iv) the correlated sample is transformed back to the original scale of the data [50].

After a comparison of bivariate Gumbel, bivariate Frank, bivariate clayton, multivariate student's t, and Gaussian copula, the Gaussian copula is found to have the best performance [50]. Therefore, the Gaussian copula, as expressed in (4.24) [45], is used for modelling the correlation among load model parameters.

$$C(u_1, u_2, \ldots, u_n; \Sigma) = \Phi_\Sigma\left(\Phi^{-1}(u_1), \Phi^{-1}(u_2), \ldots, \Phi^{-1}(u_n)\right) \tag{4.24}$$

where C represents the copula function, Σ is a symmetric, positive definite correlation matrix and $diag(\Sigma) = 1$. Φ_Σ represents the standard multivariate normal distribution, and Φ^{-1} refers to the inverse of the Normal *cdf*.

The four stages mentioned above are shown in Fig. 4.6, which are stochastic dependence pattern of k_p and k_q of the same bus (top left), transforming the data to

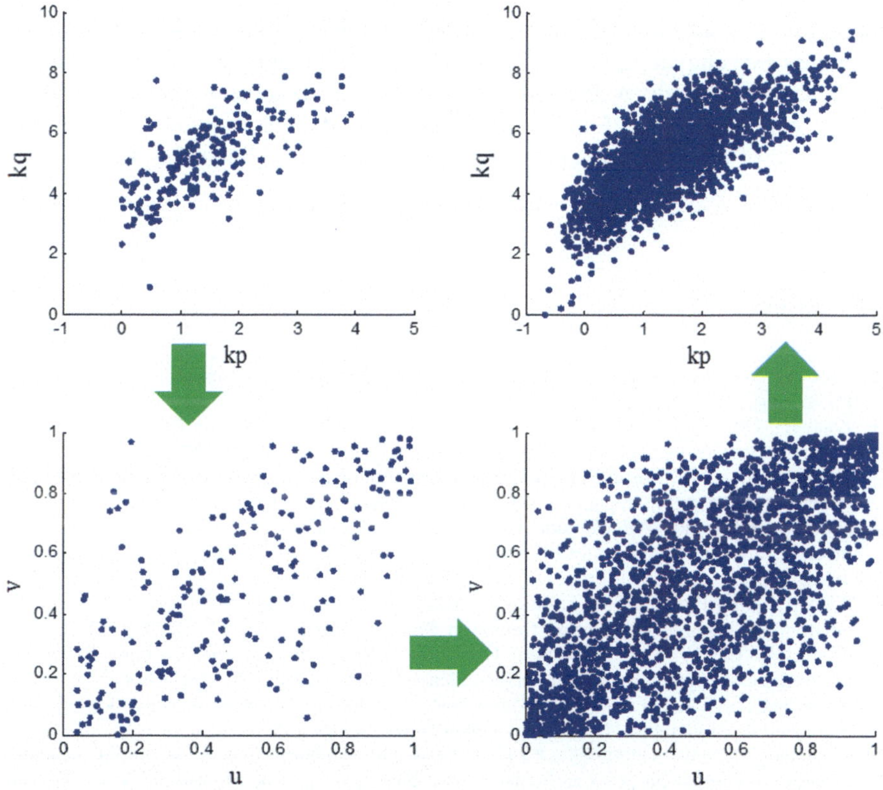

Fig. 4.6 An illustrative example of modelling stochastic dependence of uncertainties

the unit square with u representing the transformed k_p value and v representing the transformed k_q value (bottom left), using multivariate Gaussian copula to generate correlated samples (bottom right), and transforming the correlated sample back to the original scale of the data (top right). If comparing the top left and top right figure of Fig. 3.14, it is clear that the data generated by using multivariate Gaussian copula keeps the correlation pattern presented in the raw data. The benefit of generating correlated data using copula theory is that the generated samples have taken correlation among parameters into consideration, which is not considered in random number generation methods.

4.5 Summary

This chapter discussed probabilistic assessment, sensitivity analysis, and stochastic dependence of uncertain parameters, which will be applied in the following three chapters. This chapter started by introducing how to model uncertainties in the power system, e.g., wind power, solar power, and load variation. Then it introduced the stopping criteria for Monte Carlo simulation. After that, different sensitivity analysis techniques are compared and Morris screening method is chosen as the most suitable technique for this study. A detailed procedure for Morris screening method is given together with an example. Finally, some concepts and methods related to stochastic dependence of uncertain parameters have been introduced, including PCC, kernel density estimation, and copula.

References

1. Sarić AT, Stankovic AM (2005) Model uncertainty in security assessment of power systems. Power Syst IEEE Trans 20:1398–1407
2. Zhao JH, Dong ZY, Lindsay P, Wong KP (2009) Flexible transmission expansion planning with uncertainties in an electricity market. Power Syst IEEE Trans 24:479–488
3. Soroudi A, Aien M, Ehsan M (2012) A probabilistic modeling of photo voltaic modules and wind power generation impact on distribution networks. Syst J IEEE 6:254–259
4. Zou K, Agalgaonkar AP, Muttaqi KM, Perera S (2012) Distribution system planning with incorporating DG reactive capability and system uncertainties. Sustain Energy IEEE Trans 3:112–123
5. Aien M, Fotuhi-Firuzabad M, Aminifar F (2012) Probabilistic load flow in correlated uncertain environment using unscented transformation. Power Syst IEEE Trans 27:2233–2241
6. Zhang JF, Tse C, Wang K, Chung C (2009) Voltage stability analysis considering the uncertainties of dynamic load parameters. Gener Transm Distrib IET 3:941–948
7. Valverde G, Saric AT, Terzija V (2013) Stochastic monitoring of distribution networks including correlated input variables. IEEE Trans Power Deliv 28:246–255
8. Evangelopoulos VA, Georgilakis PS (2014) Optimal distributed generation placement under uncertainties based on point estimate method embedded genetic algorithm. Gener Transm Distrib IET 8:389–400
9. Ahmadi H, Ghasemi H (2012) Maximum penetration level of wind generation considering power system security limits. Gener Transm Distrib IET 6:1164–1170

10. Faried SO, Billinton R, Aboreshaid S (2009) Probabilistic evaluation of transient stability of a wind farm. Energy Convers IEEE Trans 24:733–739
11. Li Y, Li W, Yan W, Yu J, Zhao X (2014) Probabilistic optimal power flow considering correlations of wind speeds following different distributions. Power Syst IEEE Trans 29:1847–1854
12. Zhao X, Zhou J (2010) Probabilistic transient stability assessment based on distributed DSA computation tool. In: 2010 IEEE 11th international conference on probabilistic methods applied to power systems (PMAPS). pp 685–690
13. Chiodo E, Gagliardi F, La Scala M, Lauria D (1999) Probabilistic on-line transient stability analysis. In: IEEE proceedings on generation, transmission and distribution. pp 176–180
14. Helton JC, Davis FJ (2003) Latin hypercube sampling and the propagation of uncertainty in analyses of complex systems. Reliab Eng Syst Saf 81:23–69
15. Fishman G (2013) Monte Carlo: concepts, algorithms, and applications. Springer Science & Business Media
16. Preece R, Huang K, Milanovic JV (2014) Probabilistic small-disturbance stability assessment of uncertain power systems using efficient estimation methods. Power Syst IEEE Trans 29:2509–2517
17. Valenzuela J, Mazumdar M (2001) Monte Carlo computation of power generation production costs under operating constraints. Power Syst IEEE Trans 16:671–677
18. El-Khattam W, Hegazy Y, Salama M (2006) Investigating distributed generation systems performance using Monte Carlo simulation. IEEE Trans Power Syst 21:524–532
19. Martinez JA, Guerra G (2014) A parallel Monte Carlo method for optimum allocation of distributed generation. Power Syst IEEE Trans 29:2926–2933
20. Midence D (2008) Distribution system planning considering the social balance of supply quality. Ph.D. Dissertation. Universidad Nacional de San Juan. Facultad de Ingeniería. Instituto de Energía Eléctrica
21. Rueda JL, Colomé DG, Erlich I (2009) Assessment and enhancement of small signal stability considering uncertainties. Power Syst IEEE Trans 24:198–207
22. Saltelli A, Chan K, Scott EM (2000) Sensitivity analysis, vol 1. Wiley, New York
23. Song X, Zhang J, Zhan C, Xuan Y, Ye M, Xu C (2015) Global sensitivity analysis in hydrological modeling: review of concepts, methods, theoretical framework, and applications. J Hydrol 523:739–757
24. Iooss B, Lemaître P (2015) A review on global sensitivity analysis methods. In: Uncertainty management in simulation-optimization of complex systems. Springer, pp 101–122
25. Hasan K, Preece R, Milanović J (2016) Efficient identification of critical parameters affecting the small-disturbance stability of power systems with variable uncertainty. Power Energy Soc General Meet (PESGM) 2016:1–5
26. Ruano M, Ribes J, Ferrer J, Sin G (2011) Application of the Morris method for screening the influential parameters of fuzzy controllers applied to wastewater treatment plants. Water Sci Technol 63:2199–2206
27. Bettonvil B, Kleijnen JP (1997) Searching for important factors in simulation models with many factors: sequential bifurcation. Eur J Oper Res 96:180–194
28. Papaefthymiou G, Kurowicka D (2009) Using copulas for modeling stochastic dependence in power system uncertainty analysis. IEEE Trans Power Syst 24:40–49
29. Park H, Baldick R, Morton DP (2015) A stochastic transmission planning model with dependent load and wind forecasts. IEEE Trans Power Syst 30:3003–3011
30. Zhang N, Kang C, Singh C, Xia Q (2016) Copula based dependent discrete convolution for power system uncertainty analysis. IEEE Trans Power Syst 31:5204–5205
31. Wu W, Wang K, Han B, Li G, Jiang X, Crow ML (2015) A versatile probability model of photovoltaic generation using pair copula construction. IEEE Transactions on Sustainable Energy 6:1337–1345
32. Li P, Guan X, Wu J, Zhou X (2015) Modeling dynamic spatial correlations of geographically distributed wind farms and constructing ellipsoidal uncertainty sets for optimization-based generation scheduling. IEEE Trans Sustainable Energy 6:1594–1605

33. Haghi HV, Lotfifard S (2015) Spatiotemporal modeling of wind generation for optimal energy storage sizing. IEEE Trans Sustainable Energy 6:113–121
34. Bina MT, Ahmadi D (2015) Stochastic modeling for the next day domestic demand response applications. IEEE Trans Power Syst 30:2880–2893
35. Moriya N (2008) Noise-related multivariate optimal joint-analysis in longitudinal stochastic processes. In: Progress in applied mathematical modeling. p 223
36. Cohen J (1988) Statistical power analysis for the behavioral sciences, 2nd edn. Erlbaum, Hillsdale, NJ
37. Buda A, Jarynowski A (2010) Life time of correlations and its applications. Andrzej Buda Wydawnictwo NiezaleĹĽne
38. Parzen E (1962) On estimation of a probability density function and mode. Ann Math Stat 33:1065–1076
39. Wand MP, Jones MC (1994) Kernel smoothing. Chapman and Hall/CRC
40. Epanechnikov VA (1969) Non-parametric estimation of a multivariate probability density. Theor Probab Appl 14:153–158
41. Park BU, Marron JS (1990) Comparison of data-driven bandwidth selectors. J Am Stat Assoc 85:66–72
42. Park B, Turlach B (1992) Practical performance of several data driven bandwidth selectors. Université catholique de Louvain, Center for Operations Research and Econometrics (CORE)
43. Cao R, Cuevas A, Manteiga WG (1994) A comparative study of several smoothing methods in density estimation. Comput Stat Data Anal 17:153–176
44. Jones MC, Marron JS, Sheather SJ (1996) A brief survey of bandwidth selection for density estimation. J Am Stat Assoc 91:401–407
45. Hentati R, Prigent J-L (2010) Chapter 4 Copula theory applied to hedge funds dependence structure determination. In: Nonlinear modeling of economic and financial time-series. Emerald Group Publishing Limited, pp 83–109
46. Cai D, Shi D, Chen J (2014) Probabilistic load flow computation using Copula and Latin hypercube sampling. IET Gener Transm Distrib 8:1539–1549
47. Saadat N, Choi SS, Vilathgamuwa DM (2015) A statistical evaluation of the capability of distributed renewable generator-energy-storage system in providing load low-voltage ride-through. IEEE Trans Power Delivery 30:1128–1136
48. Zhang N, Kang C, Xia Q, Liang J (2014) Modeling conditional forecast error for wind power in generation scheduling. IEEE Trans Power Syst 29:1316–1324
49. Park H, Baldick R (2015) Stochastic generation capacity expansion planning reducing greenhouse gas emissions. IEEE Trans Power Syst 30:1026–1034
50. Hasan KN, Preece R (2017) Influence of stochastic dependence on small-disturbance stability and ranking uncertainties. IEEE Trans Power Syst

Chapter 5
Load Model Parameter Ranking for Different Types of Power System Stability Studies

In this chapter, the Morris screening method is applied for identifying critical load model parameters for four types of power system stability studies. A level of 10 is chosen for each uncertain parameter, which is for a higher confidence level and not computational demanding. For instance, in order to obtain the ranking order for composite load model, the number of simulation needs to be performed is $35 \times 12 \times 10 + 1 = 4201$, where 35 is the number of load buses, 12 is the number of parameters for each load model, 10 is the number of variations for each parameter, and 1 is the base case. The ranking has been carried out for various load models and loading conditions in order to investigate their influence on critical load model parameters.

5.1 Test Network

In this study, the modified version of a reduced order equivalent model of NETS-NYPS test system (New England Test System—New York Power System) was used as the simulation network. The diagram of the system is shown in Fig. 3.8. The system has 5 areas, 16 generators, and 68 buses. Among these 68 buses, 35 buses are load buses. The generators G1-G9 belong to NETS, and G10-G13 are generators of NYPS. G14, G15, and G16 represent the other three neighbouring areas. Among these generators, G9 is equipped with a fast-acting static exciter (IEEE STIA) and power system stabilizer (PSS). Other generators use a slow exciter (IEEE DC1A). All generators are equipped with speed governor systems. Generator G1 includes a GAST speed governor. G3 and G9 comprise IEEEG3 (hydro turbine), and G2-G8, G10-G16 contain IEEEG1 (steam turbine). The synchronous generators are modelled by sixth order models. The transmission lines are represented by the standard π equivalent circuit. Detailed network parameters are included in Appendix A.

There are two types of renewable energy sources (RES) modelled in the test system, which are Type 3 doubly fed induction generators (DFIG) and Type 4 Full Converter Connected (FCC) units, as shown in Fig. 5.1 and Fig. 5.2. The Type 3 model, suitable for large scale stability studies, is similar to the model proposed by

Y. Zhu, *Power System Loads and Power System Stability*, Springer Theses,
https://doi.org/10.1007/978-3-030-37786-1_5

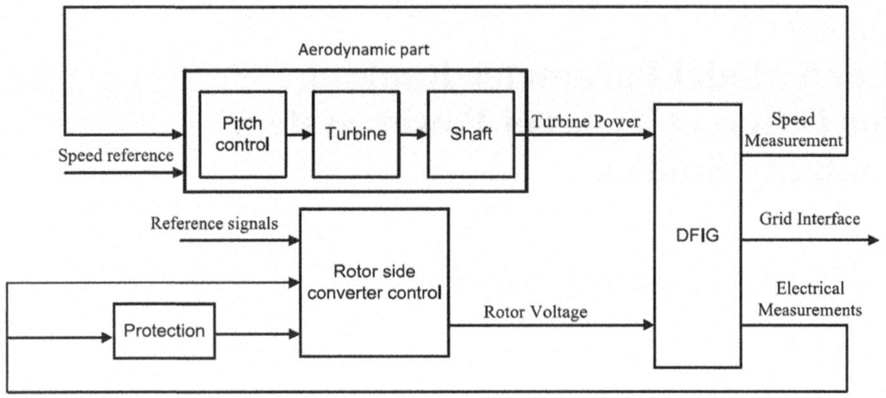

Fig. 5.1 DFIG control structure, adopted from [2]

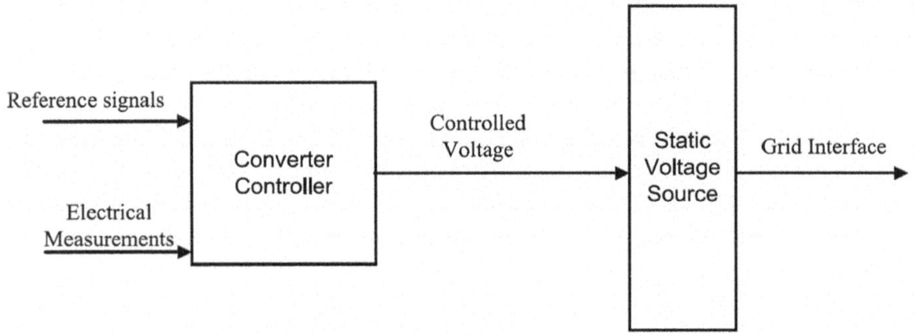

Fig. 5.2 FCC unit control structure, adopted from [2]

WECC and IEC in their structure [1]. The structure of Type 3 model is illustrated in Fig. 3.9. This model takes into consideration the pitch control of the blades, the shaft of the wind turbine and the aerodynamic part. The rotor side converter controller is also modelled including ramp rates, protection mechanisms, and relevant limitations. The DFIG is modelled by a typical 2nd order induction machine model. The rotor side converter controls the voltage in the rotor. Thus, this model considers all the relevant parts that affect the dynamic behaviour of DFIGs. The Type 3 model is used to model wind turbines. The Type 4 model can be used to model both, wind turbines and photovoltaic (PV) units in stability studies, because the converter can be considered to decouple the dynamics of the source on the dc part. The Type 4 model used in this thesis shown in Fig. 3.10 has a similar structure to the model discussed in [1]. The wind speed follows a Weibull distribution with a scale parameter of 11.1 and shape parameter of 2.2 while the sun light follows a beta distribution with a = 13.7 and b = 1.3 [2].

5.2 Procedure for Parameter Ranking

The procedure for load model parameter ranking is illustrated in Fig. 5.3. It contains three stages. Stage I and Stage III are run in MATLAB R2013a, and Stage II is run in DIgSILENT PowerFactory 2017 SP1.

The purpose of Stage I is generating load model and other power system parameters using Morris screening method. The first step is choosing the type of power

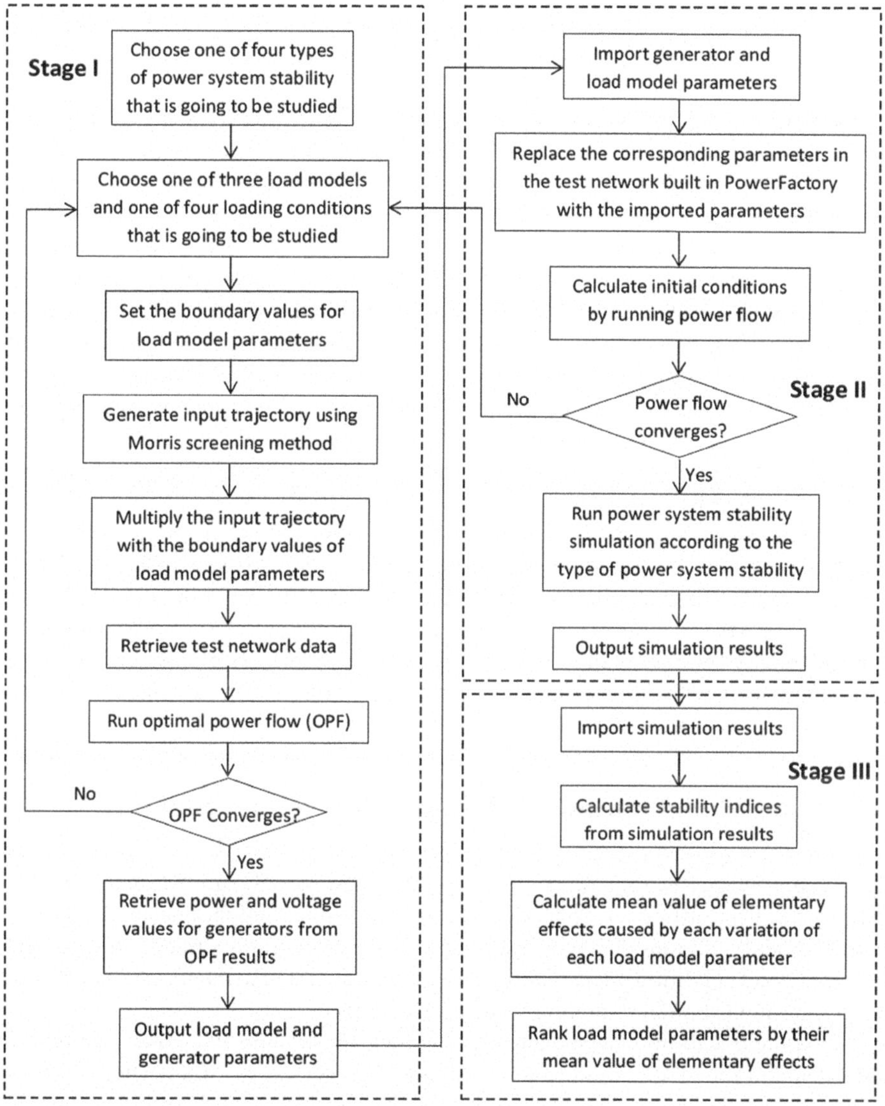

Fig. 5.3 Load model parameter ranking procedure

system stability study. Here the transient stability is used as an example. Next step is choosing the load model that is going to be studied, for example, static exponential load model. Also, the loading level has to be set. In this study, the maximum and minimum loading levels are found by increasing or decreasing loading level while running OPF until the OPF does not converge. For next step, a random input trajectory is generated for load model parameters. The details are explained in Sect. 4.3.2. Then the load model parameters are multiplied with this trajectory to obtain random load model parameters. After the test network data are retrieved, OPF is run. If the OPF does not converge, then the load model and loading conditions need to be adjusted. After OPF has been successfully performed, the generator power voltage can be determined. Finally, MATLAB output all parameters into text files for Stage II to use.

The purpose of Stage II is performing power system stability analysis using the parameters obtained in Stage I. Firstly, the DIgSIELNT PowerFactory opens and reads the files generated in Stage I. Then the values obtained are used to replace the parameter values in the test network model built in DIgSILENT PowerFactory. After that, a power flow is performed to calculate the initial conditions. This power flow may not converge because the DIgSIELTNT PowerFactory has a more detailed test network model than in MATLAB, as described in Sect. 5.4.1, which has more constraints. If the power flow does not converge, then the procedure needs to roll back to Stage I to adjust load model and loading conditions. If the initial conditions have been successfully computed, then the RMS simulation will be run. A same fault has been applied to each simulation, and the changes of rotor angle values of all generators are recorded. Finally, the results of each simulation are recorded into text file for Stage III to use.

The purpose of Stage III is calculating the stability indices from the simulation results and ranking load model parameters. Firstly, MATLAB reads the files generated in Stage II. Then it calculates TSI for each simulation from the obtained rotor angle values. After that the mean of elementary effects for each load model parameter is calculated by comparing the TSI values with the input trajectory obtained in Stage I. The parameter with a higher mean of elementary effects is more critical.

5.3 Load Model Parameter Ranking for Voltage Stability

This section discusses ranking of power system load model parameters for voltage stability for three different load models (static exponential, polynomial, and composite load model) and four different loading conditions (maximum, rated, average, and minimum load condition). The ranking results are illustrated in the form of heatmaps, and discussions about the ranking results are provided.

The voltage stability assessment is performed by running PV curve calculation and calculating the PV margin from the obtained power values of the whole system.

5.3.1 Parameter Ranking for Different Load Models

The load model parameter rankings for voltage stability are presented in Figs. 5.4, 5.5 and 5.6 for static exponential load model, polynomial load model, and composite load model respectively. They are in the form of heatmap, where the parameter marked with a darker colour is more influential. The ranking orders are calculated by (4.9). There are 35 load buses (L1–L52), and each load has 4 parameters for static exponential load model, 8 parameters for polynomial load model, and 12 parameters for composite load model.

From Fig. 5.4, it can be seen that the parameter importance order is k_p, k_q, P_0, and Q_0. The k_p parameters at Buses 17, 18, 39, 41, and 42 are the most critical parameters among all load model parameters at all buses. All parameters of these important buses tend to be more influential than the same parameters of other buses. In Fig. 5.5, the important load buses are basically the same compared with Fig. 5.6. Active power parameters P_0, p_1, p_2, and p_3 are still more critical than the reactive power parameters Q_0, q_1, q_2, and q_3. The voltage exponents are also more influential than load size parameters. Because the sum of the effects of p_1, p_2, and p_3 is larger than P_0, and the sum of the effects of q_1, q_2, and q_3 is larger than Q_0. The cumulative influence of p_1, p_2, and p_3 is equivalent to the influence of k_p, and the same is true for reactive power parameters. For instance, if p_1 has much larger value than p_2 and p_3, then in the static exponential load model, k_p will be close to 2. If p_3 is larger than p_1

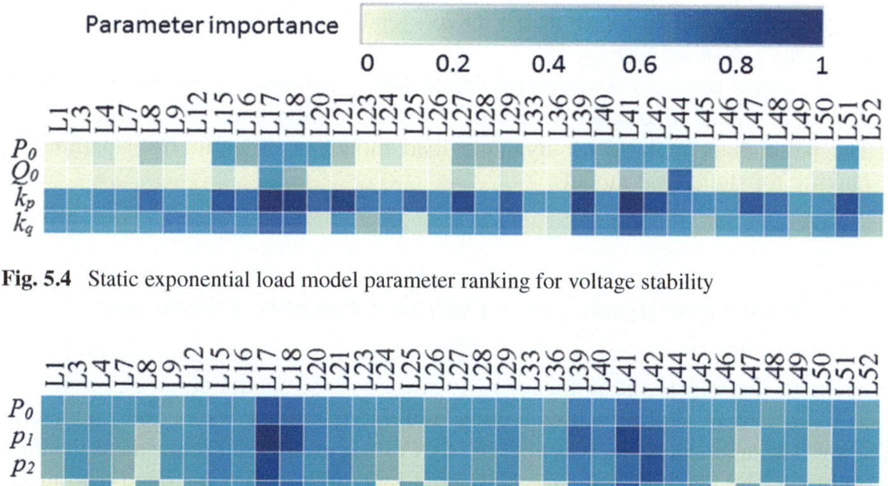

Fig. 5.4 Static exponential load model parameter ranking for voltage stability

Fig. 5.5 Polynomial load model parameter ranking for voltage stability

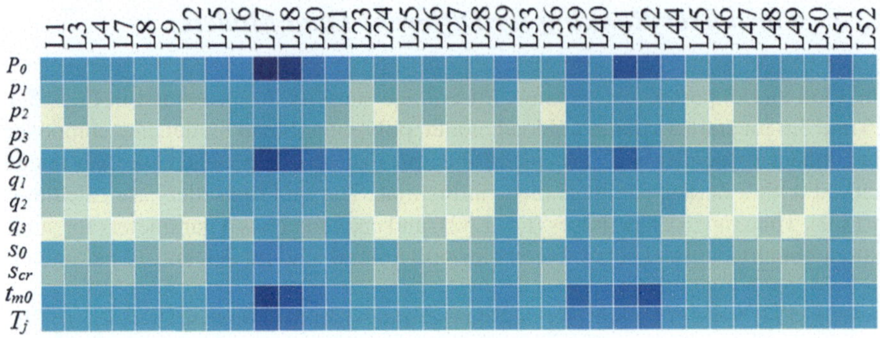

Fig. 5.6 Composite load model parameter ranking for voltage stability

and p_2, then in the static exponential load model, k_p will be close to 0. All parameters of important buses are still more critical than the same parameters of other buses. In Fig. 5.6, the load size parameters P_0 and Q_0 become more influential than they are in static exponential load model and polynomial load model. Because P_0 and Q_0 not only represent the size of the static load, but also the size of the dynamic load. Active power parameters are still more critical than reactive power parameters, although the difference is not that obvious now. The order of the important parameters is P_0, Q_0, t_{m0}, T_j, and s_0.

After comparing Figs. 5.4, 5.5 and 5.6, the following conclusions can be made:

- For important load buses, all its parameters tend to be more critical than the same parameters of other buses.
- Active power parameters are more critical than reactive power parameters.
- The most critical load buses remain unchanged for different load models.
- The parameters ranking for dynamic load model is different from parameter ranking for static load model.

5.3.2 *Parameter Ranking for Different Loading Conditions*

In order to investigate the influence of loading on load model parameter ranking, the power system loading varies among minimum, average, rated and maximum loading. The annual loading curve of NETS-NYPS is used to determine these loading conditions. The loading factor for maximum loading is 1.318, for rated loading is 1, for average loading is 0.495, and for minimum loading is 0.29. In real power system, the loading changes constantly and can be any value on the annual loading curve. However, the method of ranking load model parameters is not for online applications because it is neither necessary nor feasible to obtain a set of load model parameter ranking at each moment. These four loading levels are chosen to represent characteristic loading conditions so that a robust ranking order can be obtained. The

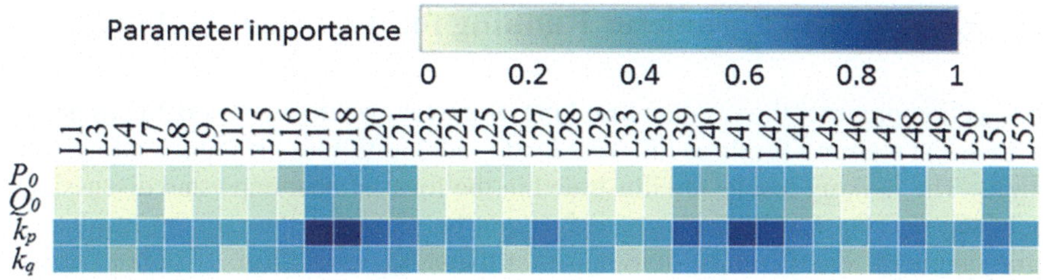

Fig. 5.7 Static exponential load model parameter ranking for voltage stability at maximum loading

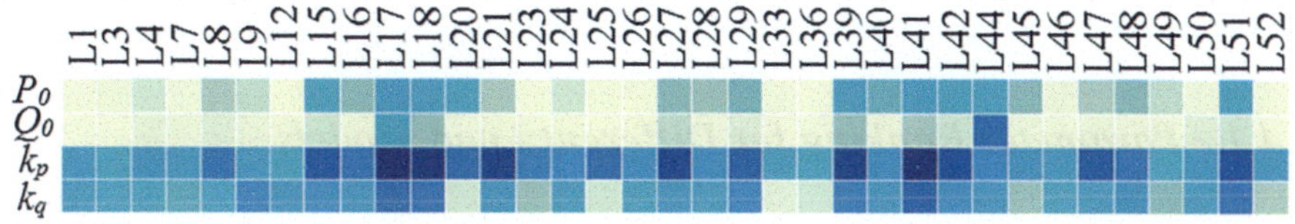

Fig. 5.8 Static exponential load model parameter ranking for voltage stability at rated loading

Fig. 5.9 Static exponential load model parameter ranking for voltage stability at average loading

Fig. 5.10 Static exponential load model parameter ranking for voltage stability at minimum loading

load model is static exponential load model. The parameter rankings for maximum, rated, average, and minimum loading are shown in Figs. 5.7, 5.8, 5.9 and 5.10.

The ranking results show that the ranking order is similar between high loading conditions (maximum and rated), and between low loading conditions (average and minimum), but the difference between high and low loading conditions is obvious. For different loading conditions, the top three loads remain the same but the 4th and 5th load changes as the loading condition varies. It can also be seen that at lower loading conditions, the load size parameters P_0 and Q_0 become more influential.

5.4 Load Model Parameter Ranking for Transient Stability

This section presents the results of ranking of power system load model parameters for transient stability for three different load models and four different loading conditions. The ranking results are illustrated in the form of heatmaps, and discussions about the ranking results are provided. It can be seen that the ranking results are significantly different compared to those for the voltage stability.

The transient stability assessment is performed by introducing a three phase short circuit fault located at line 51 and running RMS simulation for 20 s after the fault, then calculating TSI from obtained rotor angles of all generators.

5.4.1 Parameter Ranking for Different Load Models

The load model parameter ranking results for transient stability are presented in Figs. 5.11, 5.12 and 5.13 also as heat maps.

From Fig. 5.11, it can be seen that the load size parameters P_0 and Q_0 are more influential than the voltage exponential parameters k_p and k_q. It can also be seen that active power parameters are more critical than reactive power parameters. The most critical load buses observed from Fig. 5.11 are Buses 17, 18, 41, 20, and 42. Figure 5.12 shows that the two trends mentioned above still exist. The most critical parameters remain the same, but with different order. The ranking order is Buses 17, 41, 18, 20, and 42. It can be seen from Fig. 5.13 that the most critical parameters P_0, s_{cr}, t_{m0}, T_j, and Q_0 change significantly The induction motor parameters are more influential than the static load parameters. The most critical buses remain the same, but the order changes to Buses 17, 18, 41, 42, and 20.

By comparing Figs. 5.11, 5.12 and 5.13, the ranking results can be summarised as follows:

- The most critical parameter is P_0 regardless of the load model used.
- The most critical buses remain the same for different load models, though the ranking orders may change. This is due to the facts that these are large loads, and P_0 is always the most important parameter.

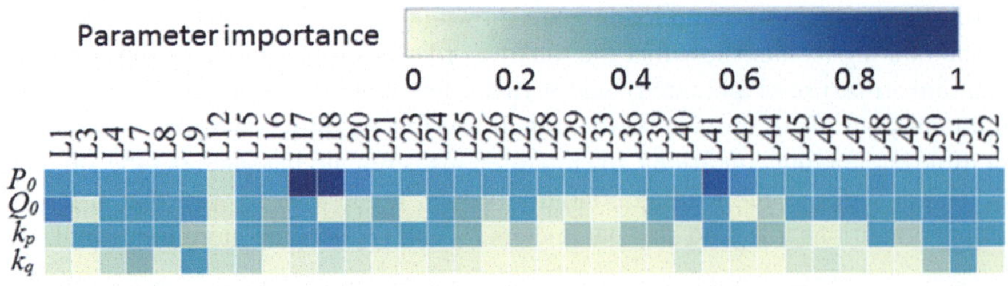

Fig. 5.11 Static exponential load model parameter ranking for transient stability

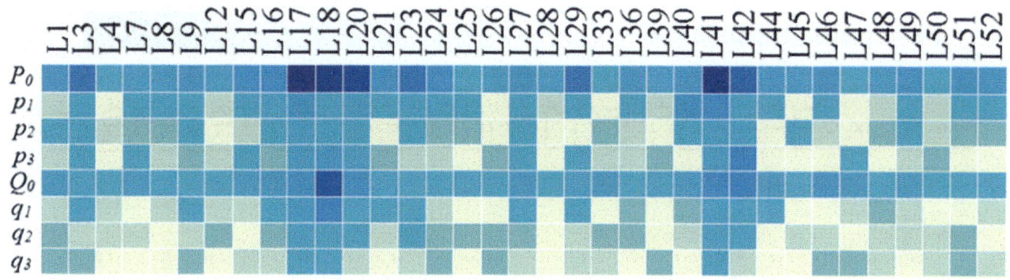

Fig. 5.12 Polynomial load model parameter ranking for transient stability

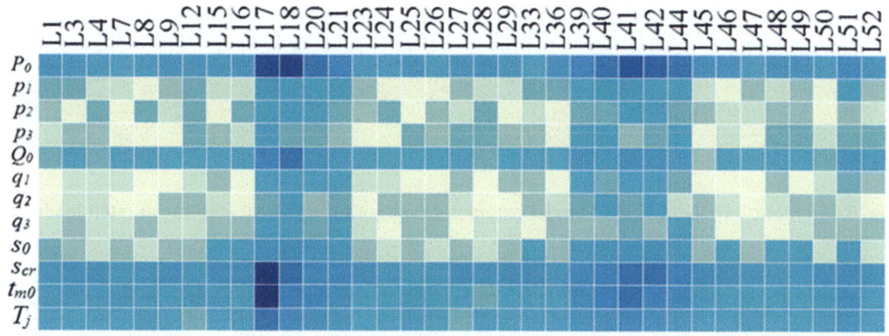

Fig. 5.13 Composite load model parameter ranking for transient stability

- For those critical load buses, all their parameters are more influential than the same parameter of other buses.
- For composite load model, dynamic load model parameters are more critical than static load model parameters.

5.4.2 *Parameter Ranking for Different Loading Conditions*

Like before, the system loading is varied from maximum loading to minimum loading in order to check the influence of loading on parameter ranking. The ranking results for maximum, rated, average, and minimum loading for transient stability are shown in Figs. 5.14, 5.15, 5.16 and 5.17.

When looking at Figs. 5.14, 5.15, 5.16 and 5.17, it can be seen that the voltage exponents are more influential at low loading conditions than high loading conditions. Most notably, the k_p parameter of Buses 18, 41, and 42. The importance of Bus 17 has reduced as loading level decreases. At high loading conditions, it is at the first place; however, at low loading conditions, it drops to fourth place. By contrast, the Bus 42 rises from fourth place in high loading conditions to the first place in low loading conditions. The four most important loads remain the same for different loading

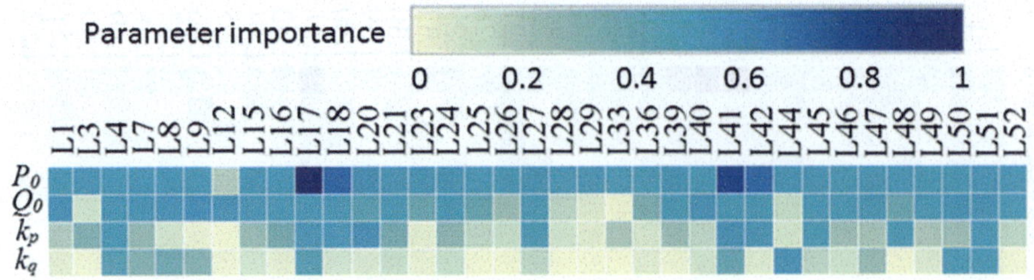

Fig. 5.14 Static exponential load model parameter ranking for transient stability at maximum loading

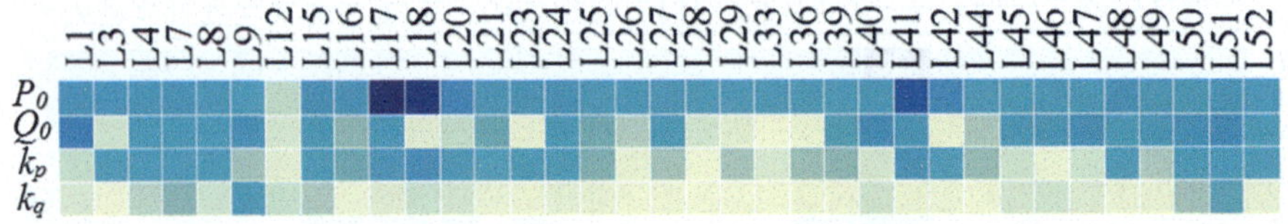

Fig. 5.15 Static exponential load model parameter ranking for transient stability at rated loading

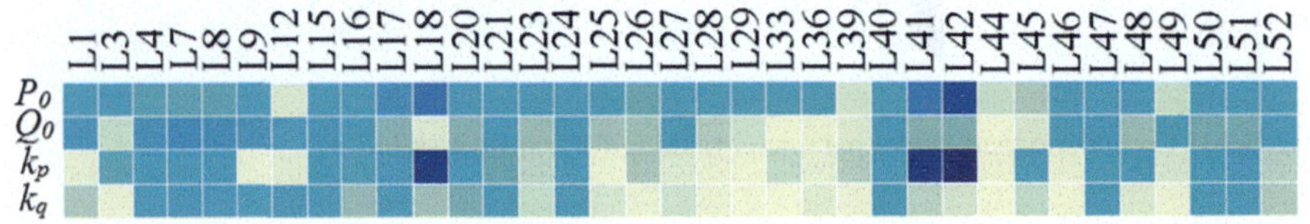

Fig. 5.16 Static exponential load model parameter ranking for transient stability at average loading

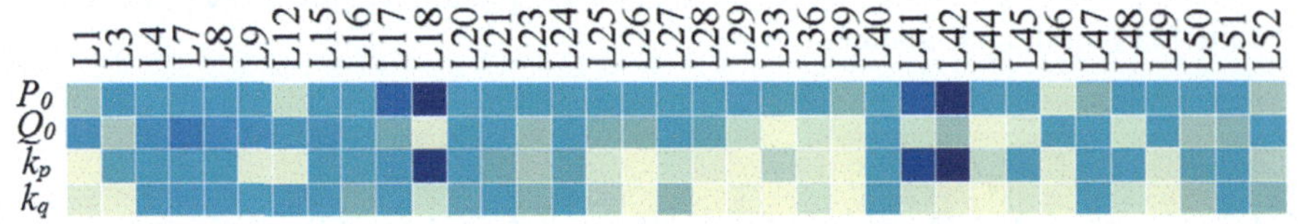

Fig. 5.17 Static exponential load model parameter ranking for transient stability at minimum loading

conditions, which are Buses 17, 18, 41, and 42, though the order of importance is different. The fifth most important bus is Bus 7 at low loading conditions and Bus 20 at high loading conditions. The reason for the changing of ranking order as loading reduces is that for lower loading, the importance of P_0 drops, but the importance of k_p raises. Load 17, as the largest load in the system, is the most influential load at high loading conditions. But at low loading conditions, when k_p is more important, load at Bus 42, with a critical k_p parameter, raises to the first place.

5.5 Load Model Parameter Ranking for Small Disturbance Stability

This section illustrates the results of ranking of power system load model parameters for small disturbance stability for three different load models and four different loading conditions. The ranking results are illustrated in the form of heatmaps, followed by the discussion and summary of ranking results. The small disturbance stability assessment is performed by running modal analysis and identifying the critical mode from all electromechanical modes.

5.5.1 Parameter Ranking for Different Load Models

The parameter ranking results of different load models for small disturbance stability are illustrated in Figs. 5.18, 5.19 and 5.20.

Figure 5.18 shows that load size parameters P_0 and Q_0 are more influential than voltage exponential parameters k_p and k_q. The active power parameters are slightly more important than reactive parameters. The most influential buses are Buses 17, 41, 18, 42, and 51. Figure 5.19 shows that load size parameters P_0 and Q_0 are more critical than p_1, p_2, p_3, q_1, q_2, and q_3, and the difference in their importance is more obvious than in static exponential load model. The most important buses remain the same. In Fig. 5.20, it can be clearly seen that load model parameters for induction motors are more critical than static parameters p_1, p_2, p_3, q_1, q_2, and q_3. The order of the importance of parameters is P_0, T_j, t_{m0}, Q_0, and s_{cr}. The most critical buses remain the same compared with static load models.

By comparing Figs. 5.1, 5.19 and 5.20, the ranking results can be summarised as follows:

- The most critical parameter is P_0 regardless of the load models used.
- The most critical buses remain the same for different load models. This is due to the facts that these are large loads, and P_0 is always the most important parameter.
- The active power parameters are generally more influential than reactive power parameters.

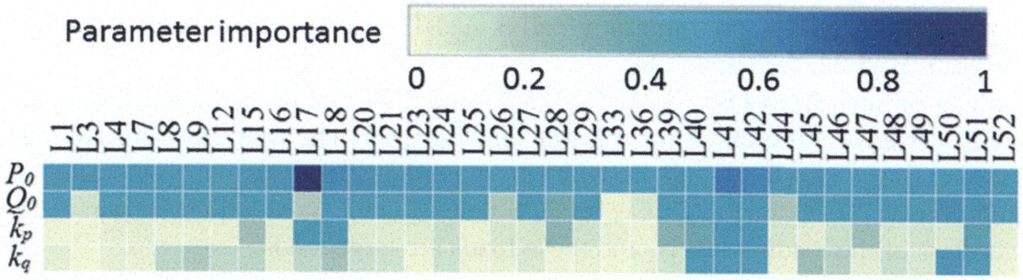

Fig. 5.18 Static exponential load model parameter ranking for small disturbance stability

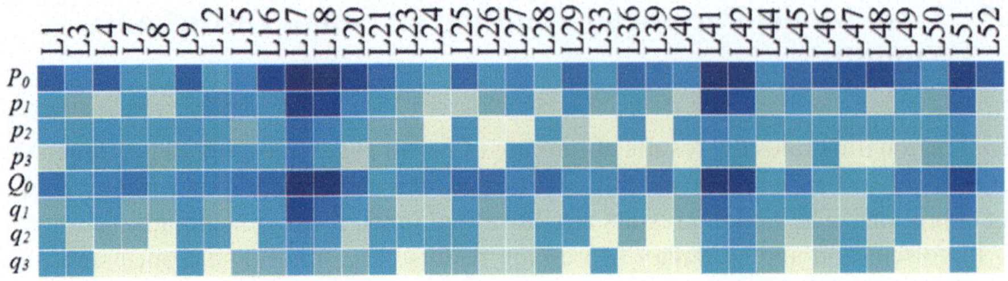

Fig. 5.19 Polynomial load model parameter ranking for small disturbance stability

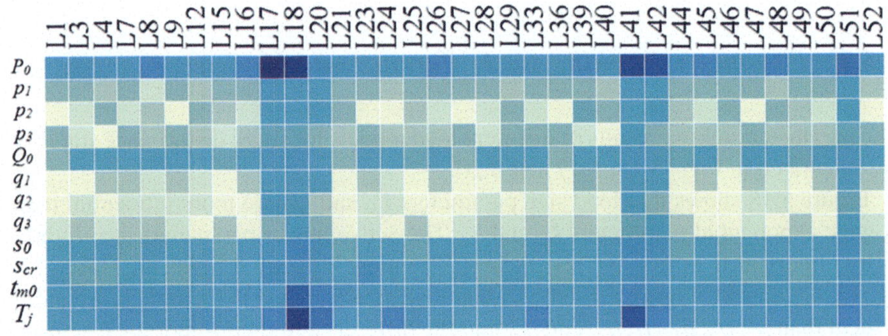

Fig. 5.20 Composite load model parameter ranking for small disturbance stability

- For those critical load buses, all their parameters are more influential than same parameter of other buses.
- For composite load model, dynamic load model parameters are more critical than static load model parameters.

5.5.2 Parameter Ranking for Different Loading Conditions

The power system loading conditions are varied as before, i.e., minimum, average, rated, and maximum loadings in order to find out the effect of loading conditions on the parameter ranking and important load buses. The ranking results are illustrated in Figs. 5.21, 5.22, 5.23 and 5.24.

It can be seen from Figs. 5.21, 5.22, 5.23 and 5.24 that the voltage exponential parameters have more effect for a lower loading level. The four most critical load buses are Buses 17, 18, 41, and 42, with varying orders. The fifth most important bus can be Buses 20, 50, or 51. This is because that P_0 is the most important parameter regardless of the loading levels, and these loads are large loads.

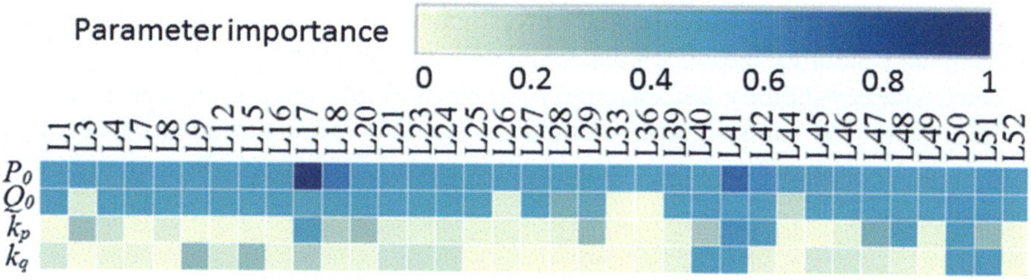

Fig. 5.21 Static exponential load model parameter ranking for small disturbance stability at maximum loading

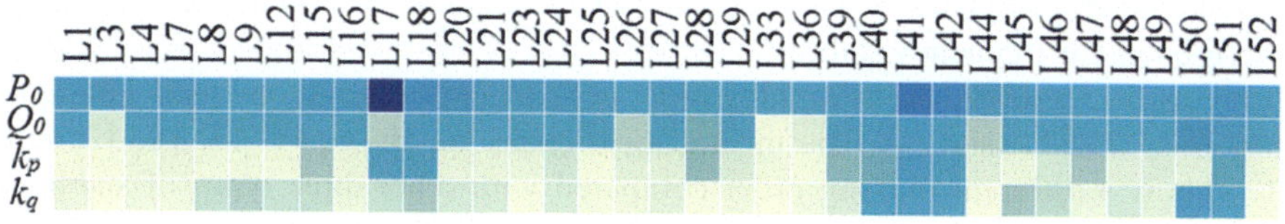

Fig. 5.22 Static exponential load model parameter ranking for small disturbance stability at rated loading

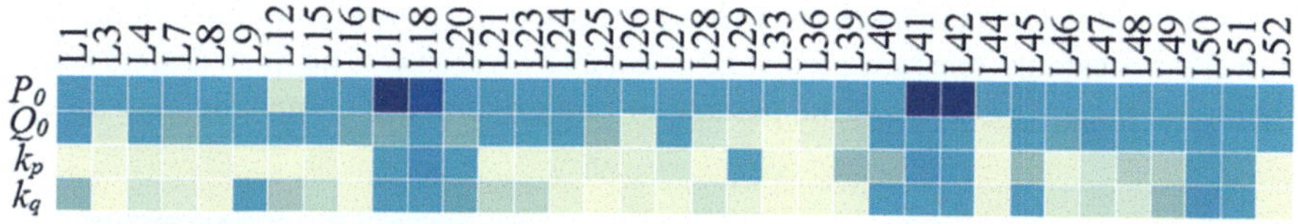

Fig. 5.23 Static exponential load model parameter ranking for small disturbance stability at average loading

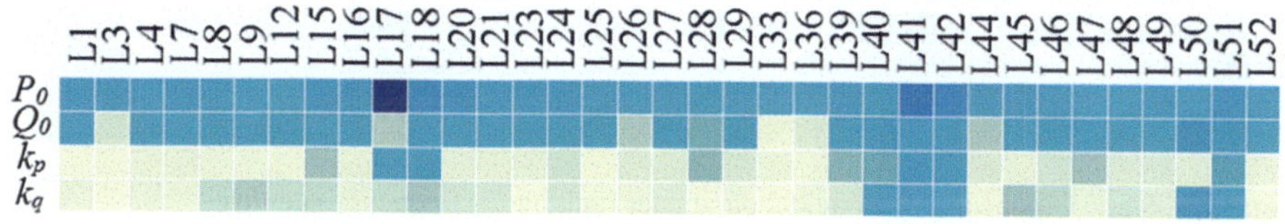

Fig. 5.24 Static exponential load model parameter ranking for small disturbance stability at minimum loading

5.6 Load Model Parameter Ranking for Frequency Stability

This section discusses ranking of power system load model parameters for frequency stability for three different load models and four different loading conditions. This is the last power system stability study investigated. The features of ranking results for frequency stability have been discussed and summarised.

The frequency stability assessment is performed by introducing a load step change in the system and running RMS simulation for 20 s after the disturbance, then identifying the lowest frequency value from obtained frequency values of all buses.

5.6.1 Parameter Ranking for Different Load Models

The last stability study is frequency stability. Figures 5.25, 5.26 and 5.27 illustrate the load model parameter ranking for static exponential, polynomial, and composite load models for frequency stability.

Figure 5.25 shows that the ranking order of static exponential load model parameters is k_p, P_0, k_q, and Q_0. The most critical buses are Buses 41, 17, 18, 42, and 51. This is a different ranking order compared with other three types of stability studies. Figure 5.26 illustrates that the ranking order following the same pattern of static exponential load model. The ranking order is p_1, P_0, p_3, q_1, Q_0, p_2, q_3, and q_2. The only difference is that not all active power parameters are more influential than reactive power parameters. Parameter p_2 is less critical than q_1 and Q0. This means the equivalent constant current load occupies less proportion of total load than equivalent constant impedance load and constant power load. The most critical buses remain the same. In Fig. 5.27, the parameter ranking order of composite load model

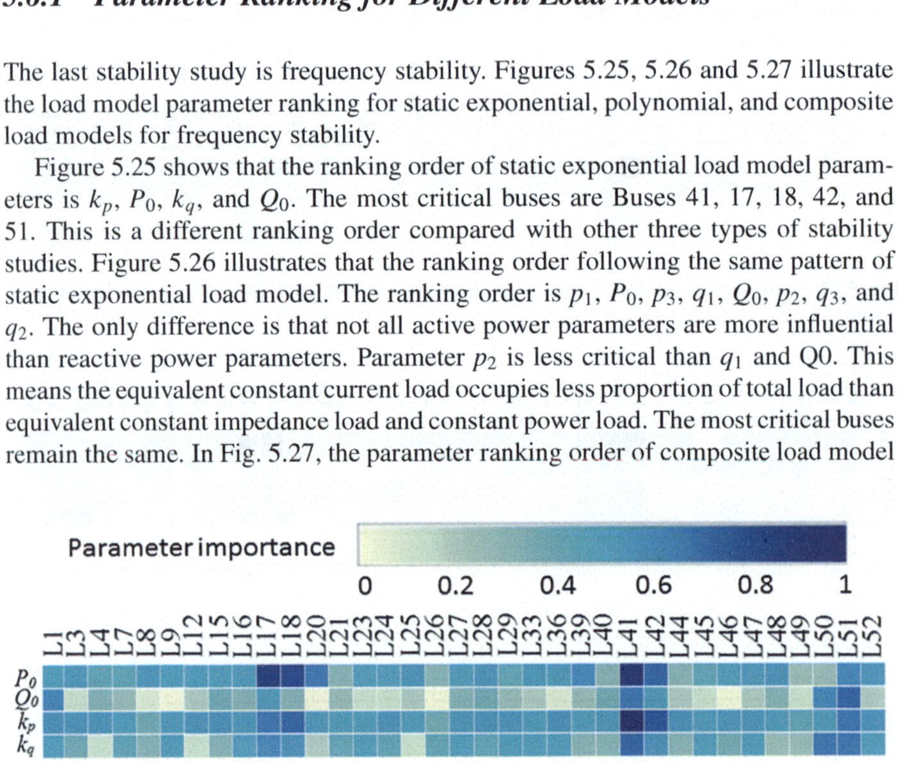

Fig. 5.25 Static exponential load model parameter ranking for frequency stability

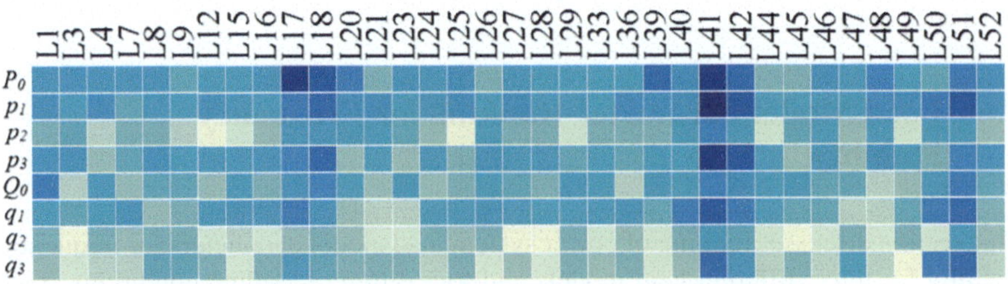

Fig. 5.26 Polynomial load model parameter ranking for frequency stability

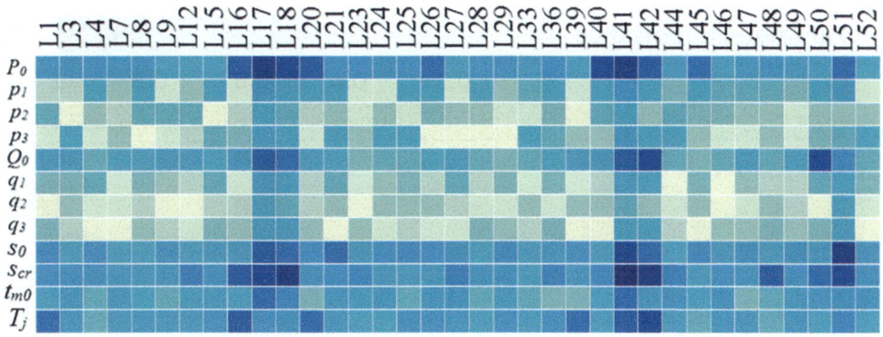

Fig. 5.27 Composite load model parameter ranking for frequency stability

is s_{cr}, s_0, P_0, T_j, t_{m0}, Q_0, p_1, p_3, q_1, p_2, q_3, and q_2. The critical slip and nominal slip are the most critical parameters for frequency stability. The most influential buses have not changed.

By comparing Figs. 5.25, 5.26 and 5.27, the ranking results can be summarised as follows:

- The most critical buses remain the same for different load models.
- The active power parameters are generally more influential than reactive power parameters.
- For those critical load buses, all their parameters are more influential than same parameter of other buses.
- For composite load model, dynamic load model parameters are more critical than static load model parameter.

5.6.2 Parameter Ranking for Different Loading Conditions

The static exponential load model parameter ranking for frequency stability at various loading conditions are shown in Figs. 5.28, 5.29, 5.30 and 5.31.

Figures 5.28, 5.29, 5.30 and 5.31 show that as loading reduces the importance of k_p and k_q increases. At the maximum loading condition, P_0 is more influential than k_p and Q_0 is more critical than k_q. The importance of Bus 41 and 42 reduces for low loading conditions. Bus 41 is the most critical bus for rated and high loading conditions, and is second and fourth for average and minimum loading conditions. In contrast, the effects of Bus 50 and 51 increases for low loading conditions. For minimum loading, Bus 50 becomes the fifth most important bus. It was not among the top five loads before that.

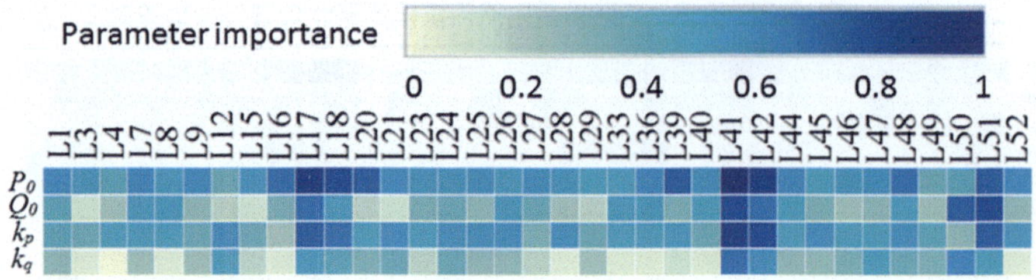

Fig. 5.28 Static exponential load model parameter ranking for frequency stability at maximum loading

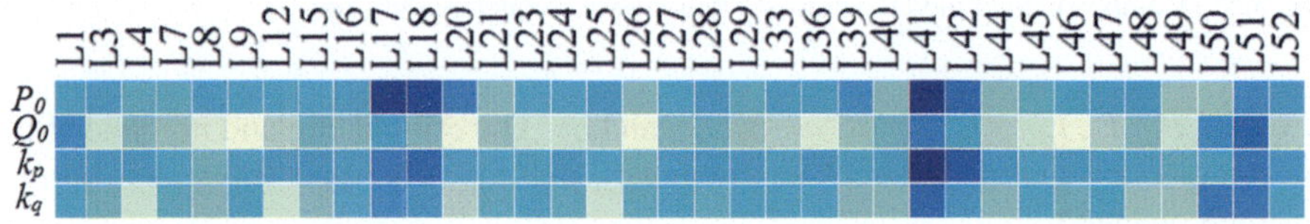

Fig. 5.29 Static exponential load model parameter ranking for frequency stability at rated loading

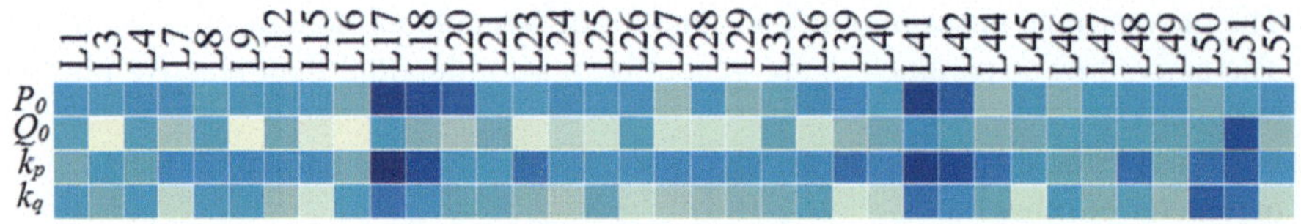

Fig. 5.30 Static exponential load model parameter ranking for frequency stability at average loading

Fig. 5.31 Static exponential load model parameter ranking for frequency stability at minimum loading

5.7 Summary of Load Model Parameter Ranking

This subsection summarises the ranking results presented in Sects. 5.2–5.5. For each type of power system stability, three different load models—static exponential load model, polynomial load model, and composite load model are applied for the parameter rankings. Some general patterns can be found from these rankings:

- For important load buses, all its parameters tend to be more critical than the same parameter on other buses.

- Active power parameters are generally more critical than reactive power parameter.
- The most critical buses generally remain unchanged for different load models and stability studies, though the ranking orders may change. This is due to the fact that these are large loads, and P_0 is usually the most important parameter.

Regardless of the load model used, the top five buses always include Bus 17, 18, 41, and 42, and they are the largest loads in the test system. This means that large loads are often critical loads for all types of stability. Sometimes for static load models, voltage exponents and proportional parameters may be more influential than load size parameters, but when dynamic load model is used, load size parameters are always more important than other static parameters. This is because that load size parameters not only reflect static loads, but also induction motors. The ranking order of induction motor parameters s_{cr}, s_0, t_{m0}, and T_j has large variations for different stability studies, but they are all more influential than static parameters p_1, p_2, p_3, q_1, q_2, and q_3.

Table 5.1 summarises the most critical load buses for four different loading conditions and four different stability studies. Table 5.1 shows that Bus 17, 18, 41, and 42 appear in the top five most often. This is due to the fact that these are largest loads in the system, and the load size parameter, P_0, is often the most critical load model parameter. Their appearance rates are 100, 94, 100, and 88%. Some other buses that appear in top five for several times are Bus 39, 48, 50, and 51. This is affected by both their load size and locations, which will be further discussed in Sect. 6.5. Bus 17 is the most important load bus for 69% of all cases, because it is by far the largest load

Table 5.1 Summary of critical loads

Ranking order		1	2	3	4	5
Voltage stability	Maximum	17	18	41	42	39
	Rated	17	18	41	42	39
	Average	17	18	41	42	48
	Minimum	17	18	41	48	51
Transient stability	Maximum	17	42	41	18	51
	Rated	17	18	41	42	51
	Average	42	41	18	17	7
	Minimum	42	18	41	17	7
Small disturbance stability	Maximum	41	17	42	18	50
	Rated	17	41	42	18	50
	Average	17	41	18	42	20
	Minimum	17	42	18	41	51
Frequency stability	Maximum	41	42	17	51	18
	Rated	41	17	18	42	51
	Average	17	41	51	18	42
	Minimum	17	51	50	41	18

in the system. For four types of stability studies, the ranking for voltage stability is most stable under different loading conditions. There are usually obvious difference in ranking between high (maximum and rated) loading conditions and low (average and minimum) loading conditions.

References

1. Göksu Ö, Sørensen PE, Morales A, Weigel S, Fortmann J, Pourbeik P (2016) Compatibility of IEC 61400-27-1 and WECC 2nd generation wind turbine models. In: 15th international workshop on large-scale integration of wind power into power systems as well as on transmission networks for offshore wind power plants
2. Papadopoulos PN, Milanović JV (2017) Probabilistic framework for transient stability assessment of power systems with high penetration of renewable generation. IEEE Trans Power Syst 32:3078–3088

Chapter 6
Factors Affecting Load Model Parameter Ranking

6.1 Introduction

Chapter 5 has identified critical load model parameters for different types of power system stability studies. This chapter looks into factors affecting the ranking of loads. The factors considered in this chapter are load size, stochastic dependence of load model parameters, load model type and load locations. The influences of these factors have been studied for four different types of system stability studies. In particular, since the importance of load locations is independent of the load models, a ranking of critical load locations has been performed, which uses a different method compared to the one presented in Chap. 5.

6.2 Effect of Load Size on Parameter Ranking

The ranking results in Chap. 5 clearly show that large loads are generally more influential, which is true for all types of stability studies. The four out of five most frequently identified buses as the most influential, Buses 17, 18, 41, and 42 are the largest loads in the test system. Parameter P_0 (effectively the size of the load) is often the most critical parameter, which also explains why large loads are often more influential. This section will provide an insight into the relation between the load size and the influence on power system stability.

Figure 6.1 shows a group of scatter plots reflecting the relation between the load size and the effect of loads on four different types of power system stability. The effect of loads are calculated by (4.9) of Morris screening method. The horizontal axis is the load size which is represented by the value of load size parameter P_0, and the vertical axis is the sum of effects of all parameters of each bus.

Figure 6.1 shows that there is nearly a linear relationship between the load effects and the logarithm of load size for large loads (>330 MW). This relationship is most obvious for transient stability, least obvious for frequency stability. However, for

Y. Zhu, *Power System Loads and Power System Stability*, Springer Theses,
https://doi.org/10.1007/978-3-030-37786-1_6

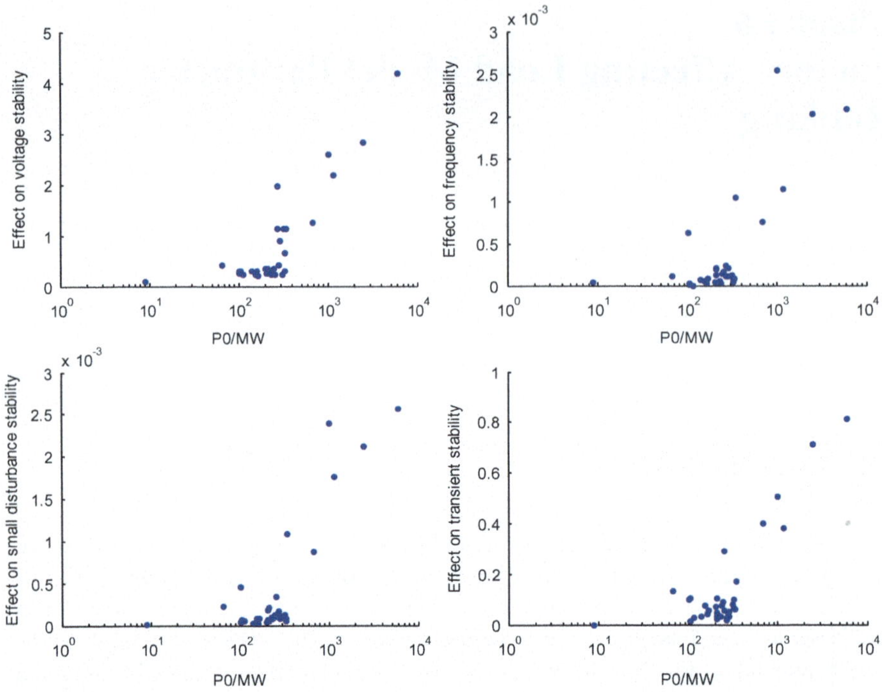

Fig. 6.1 Relation between the load size and the effect of loads on four different types of stability

small size loads, there is not a clear relationship between load size and load effects. Figure 6.1 is consistent with the results obtained in Chap. 5, where the most important loads are often large loads. These results show that load size has large influence on the load effects for all stability types. But it has to be noted that the proportional relationship between the logarithm of load size and load effects can only apply to 6 loads among 35 loads in the system. These 6 loads are significantly larger than other loads, with the largest (load 17) being 6000 MW. Most loads (27 loads) are within the range 100–330 MW. Their effects are insignificant compared to 6 large loads.

The effects of load size can also be reflected on the smallest load. The smallest load (load 12) is 9 MW, and the effects of load 12 on four types of stability are among the smallest. There is an exception in the relationship between load size and load effects of large loads, which is load 41 and 42. Load 42 (1150 MW) is slightly larger than load 41 (1000 MW), but load 41 is always more influential than load 42, sometimes even more influential than load 18 (2470 MW). This is due to the locations of load 41 and 42, which will be discussed in Sect. 6.5. The results in Fig. 6.1 show that when the difference in load size is significant, the larger load would be more influential than the smaller load.

6.3 Effect of Stochastic Dependence of Load Model Parameters

In this section, the correlation between load model parameters has been taken into consideration. The static exponential load model is used for this study because it has a simple form and a few parameters and thus has a clear correlation pattern. For complex load models, multiple parameters may be correlated with each other, which makes the analysis more complex. The correlation pattern is shown in Fig. 4.5, which is obtained from the field measurement lasted for one year. The correlation coefficients are listed in Table 6.1. Most of them can be considered as moderate relationship.

In this section, the correlation between load model parameters has been taken into consideration. The correlation pattern is shown in Fig. 4.5, which is obtained from the field measurement lasted for one year. The correlation coefficients are listed in Table 6.1. Most of them can be considered as moderate relationship.

Table 6.1 Correlation coefficients of bus 1–20

Bus number	Correlation coefficients	Bus number	Correlation coefficients
1	0.387	11	0.338
2	0.371	12	0.657
3	0.475	13	0.333
4	0.286	14	0.324
5	0.275	15	0.276
6	0.364	16	0.395
7	0.276	17	0.454
8	0.431	18	0.314
9	0.394	19	0.353
10	0.368	20	0.409

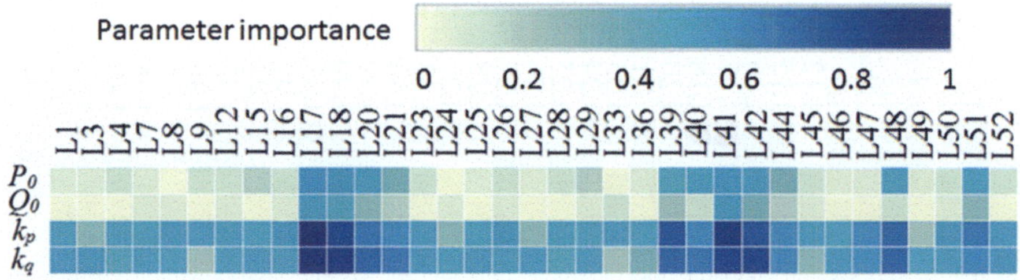

Fig. 6.2 Static exponential load model parameter ranking for voltage stability considering correlation

Fig. 6.3 Static exponential load model parameter ranking for voltage stability without correlation

The obtained ranking results for voltage stability considering correlation is shown in Fig. 6.2. For the ease of comparison, the ranking result without the consideration of correlation is reproduced as Fig. 6.3.

By comparing the ranking illustrated in Figs. 6.2 and 6.3, it can be seen that parameter k_q has become more influential when the correlation is taken into consideration. After the correlation has been considered, k_q becomes as influential as k_p. The ranking order of load size parameters has not changed because only the correlation of voltage exponent parameters is considered. Figure 4.5 shows that voltage exponents only have significant intra-correlations, i.e. correlation of parameters of the same bus, and no noticeable inter-correlations, i.e. correlation of parameters of different buses, thus the ranking order of buses is not affected.

The ranking for transient stability considering correlation is illustrated in Fig. 6.4, and the ranking result without the correlation is reproduced in Fig. 6.5.

Comparing Figs. 6.4 and 6.5 it can be seen that after taking correlation into consideration, the importance of k_q has been improved. Now the importance of k_q is similar to the importance of k_p. Similarly, the ranking orders of load size parameters

Fig. 6.4 Static exponential load model parameter ranking for transient stability with correlation

Fig. 6.5 Static exponential load model parameter ranking for transient stability without correlation

and buses have not been changed because there is only inter-correlation between voltage exponents of same bus.

The ranking for small disturbance stability considering correlation is illustrated in Fig. 6.6. As a comparison, the original ranking result is reproduced in Fig. 6.7.

It can be seen that after considering correlation, the importance of k_p and k_q have increased. Because the two parameters are correlated with each other, the variation of one of them will cause the variation of the other, resulting in a larger effect on small disturbance stability. Also, due to their correlation, the k_p and k_q on the same bus have similar ranking. In comparison, there is no obvious change in ranking for P_0 and Q_0, because there are no correlations between these parameters. The important buses also stay the same because there is only inter-correlation between voltage exponents of same bus.

Finally, the ranking of load model parameters for frequency stability after taking correlation into consideration is illustrated in Fig. 6.8 and the original ranking is reproduced in Fig. 6.9.

Comparing these two figures, it can be seen that although the change is not very obvious, the k_p and k_q on the same bus now have similar ranking. That is because they are correlated with each other, the variation of one parameter will result in the variation of the other. For P_0 and Q_0, there is no obvious change because their correlation is not modelled in this case due to lack of actual data. The critical buses stay the same due to that only inter-correlation between voltage exponents of same bus is modelled.

After this comparison, it can be concluded that the load model parameter ranking order is affected by the stochastic dependence of load model parameters. Therefore,

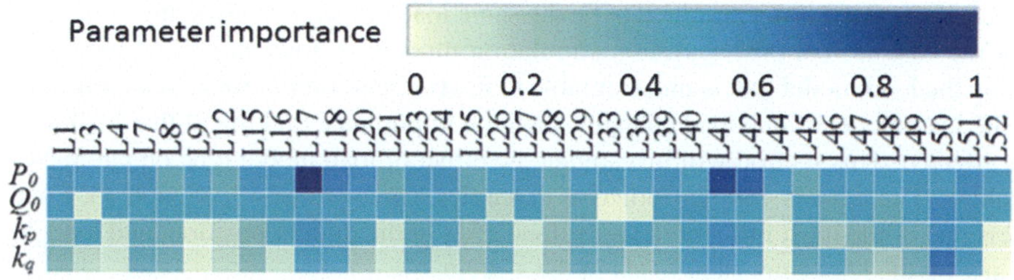

Fig. 6.6 Static exponential load model parameter ranking for small disturbance stability with correlation

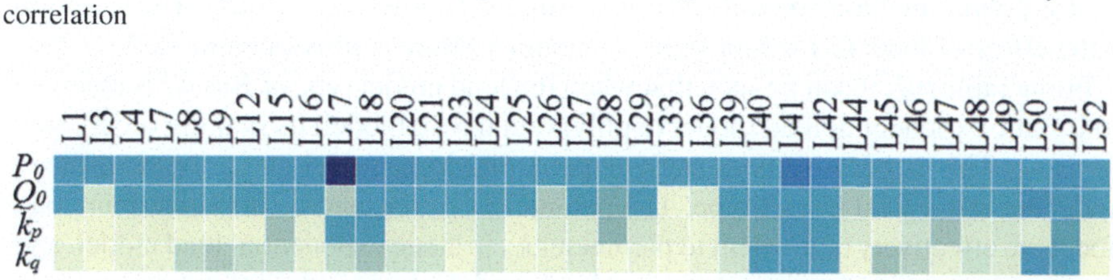

Fig. 6.7 Static exponential load model parameter ranking for small disturbance stability without correlation

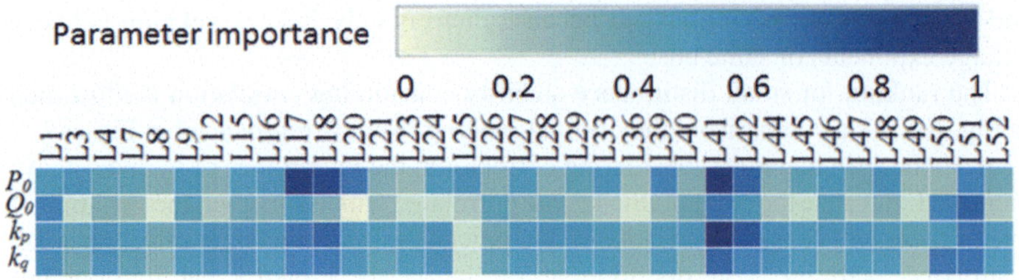

Fig. 6.8 Static exponential load model parameter ranking for frequency stability with correlation

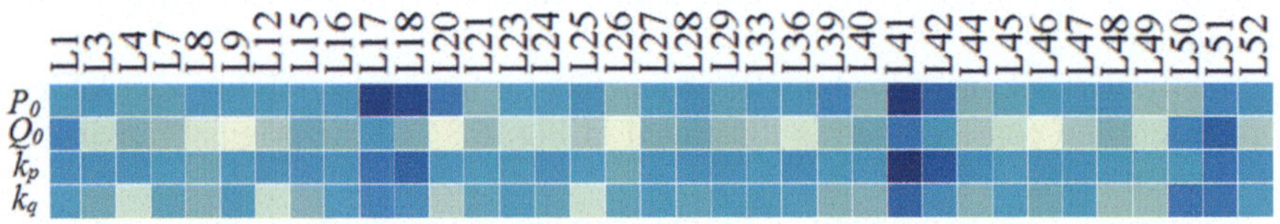

Fig. 6.9 Static exponential load model parameter ranking for frequency stability without correlation

in order to obtain the accurate load model parameter ranking, the correlation between load model parameters should be taken into consideration when it is possible.

6.4 Effect of Load Model Type on Parameter Ranking

In Chap. 5, all loads are assumed to be the same load model. However, this is usually not the case in real power system simulation models. A load should be modelled using the load model that is most suitable to it. However, under the same conditions, if the load model of one load changes, then the importance ranking of that load may be affected. This section will investigate the effect of load model type on parameter ranking. Three loads will be studied as examples, namely load 42, 15, and 33. Load 42 is one of the most influential loads, load 15 is at the middle position, and load 33 is among the least important loads in the system.

Assuming all loads are static exponential load model, except that load 42 is modelled as polynomial load model and composite load model respectively. The variation of the effect of load 42 for four types of stability study is illustrated in Table 6.2.

From Table 6.2, it can be seen that when the load model type of Bus 42 is changed from static exponential load model to polynomial load model, its ranking orders stay the same. This is because both static exponential load model and polynomial load model are static load model type. The parameters p_1, p_2, and p_3 in polynomial load model are effectively equivalent to parameter k_p in static exponential load model, and parameters q_1, q_2, and q_3 in polynomial load model are equivalent to parameter k_q in static exponential load model. Thus, changing from one load model to another one does not affect its ranking. However, when the load model is changed to composite load model, the ranking raises one position for voltage and frequency stability, and

Table 6.2 Variation of ranking of load 42

Stability types	Static exponential	Polynomial	Composite
Voltage	5	5	4
Frequency	4	4	3
Transient	5	5	3
Small disturbance	4	4	2

raises two positions for transient stability and small disturbance stability. The ranking order of Bus 42 rises because composite load model is a dynamic load model and takes induction motor into consideration. Thus its influence on the power system dynamic behaviour will be larger than when it is modelled as static load models.

The same simulations are then performed for load 15 and 33. The rankings for load 15 and 33 for four types of stability studies and three different load models are illustrated in Tables 6.3 and 6.4 respectively.

It can be seen from Table 6.3 that the variation of ranking for Bus 15 is similar to that of Bus 42, i.e., the ranking is not changed when the load model is changed from static load model to polynomial load model, and the ranking order rises when it is changed to composite load model. Compared with Bus 42, when changed to composite load model, the ranking order of Bus 15 has a larger increase, i.e., up 3 places for frequency and transient stability. Table 6.4 shows that for Bus 33, when load model is changed from static exponential load model to polynomial load model, there is a small change of ranking order, i.e., up one place except for small disturbance stability. This is probably because when the load model is changed to polynomial load

Table 6.3 Variation of ranking of load 15

Stability types	Static exponential	Polynomial	Composite
Voltage	10	10	9
Frequency	9	9	6
Transient	8	8	5
Small disturbance	9	9	7

Table 6.4 Variation of ranking of load 33

Stability types	Static exponential	Polynomial	Composite
Voltage	34	33	30
Frequency	30	29	24
Transient	33	32	26
Small disturbance	33	33	27

model, there is a small difference in load behaviour. The effects of less important loads are close to each other, thus a small difference in load model results in the change of ranking order. When the load model is changed to composite load model, there is a significant increase of ranking order (4–7 pleases). This is also because of the fact that the differences in the influence of the least important loads are small and they are grouped close to each other so the small change in effect appears to result in big change of rank.

The results show that loads modelled using dynamic load models, i.e., dynamic loads, could be more critical than loads modelled as static loads if other conditions are the same. The results also show that the ranking position of least influential loads is more sensitive to load model change. These results once again show the importance of developing accurate load models for relevant loads.

6.5 Identifying Critical Load Locations

In the power system, the load located at some location has more influence on a particular type of power system stability. Therefore, identifying the critical load locations in a power network can be beneficial for the power system operation and planning, especially for the areas where loading is increasing. Considering that because the importance of loads is also heavily affected by load size and other load model parameters, which are clearly shown in Chap. 5, it is necessary to decouple the load location from other factors affecting the influence of loads.

6.5.1 Procedure of Identifying Critical Load Locations

The Monte Carlo Simulation is used for identifying critical load locations in power system. The Morris screening method is not used because in Morris screening method, load model parameters vary in the whole range of parameter variation, thus the variation of loading is still affected by load size. Also, as shown in Sect. 6.1, the effects of loads on power system stability indices are not proportional to load size. Instead, they are more like a logarithmic relationship. The median loading of all loads in the test network is 234 MW. The variation is thus set as 3σ equals to 10% of this median loading value for each load. The uncertainties of load variation are generated and recorded in MATLAB. Then DIgSILENT PowerFactory uses these uncertainties to perform different types of power system stability analysis. The critical load locations are identified by calculating the Pearson correlation between the load variation on these locations and the variation of power system stability indices. A flow chart of this procedure is illustrated in Fig. 6.10.

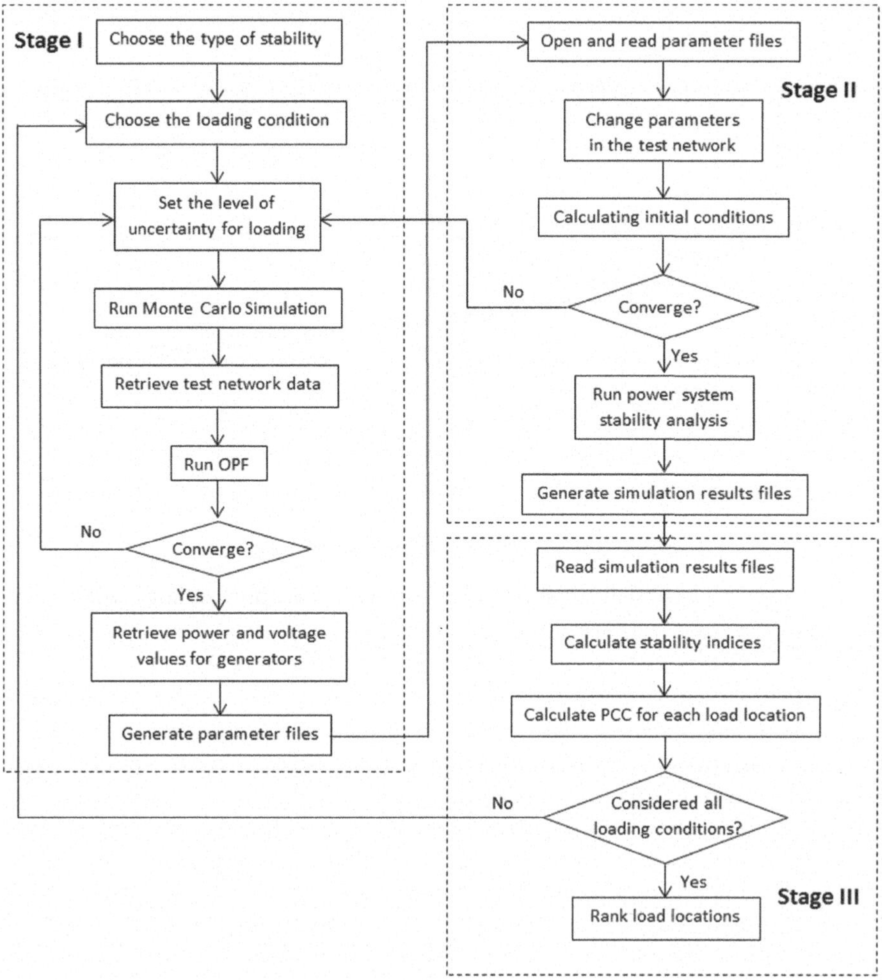

Fig. 6.10 Procedure for identifying critical load locations

6.5.2 *Annual Loading Curve*

Unlike the Morris screening method, where the load model parameters can vary within their full range, the Monte Carlo simulation can only vary the load model parameters around its nominal value with specified uncertainty $3\sigma = 10\%$. In order to compensate for this limitation, many different loading conditions have been applied. The purpose is to represent all possible loading conditions. These loading conditions are chosen from the annual system loading curve. The annual loading curve is a

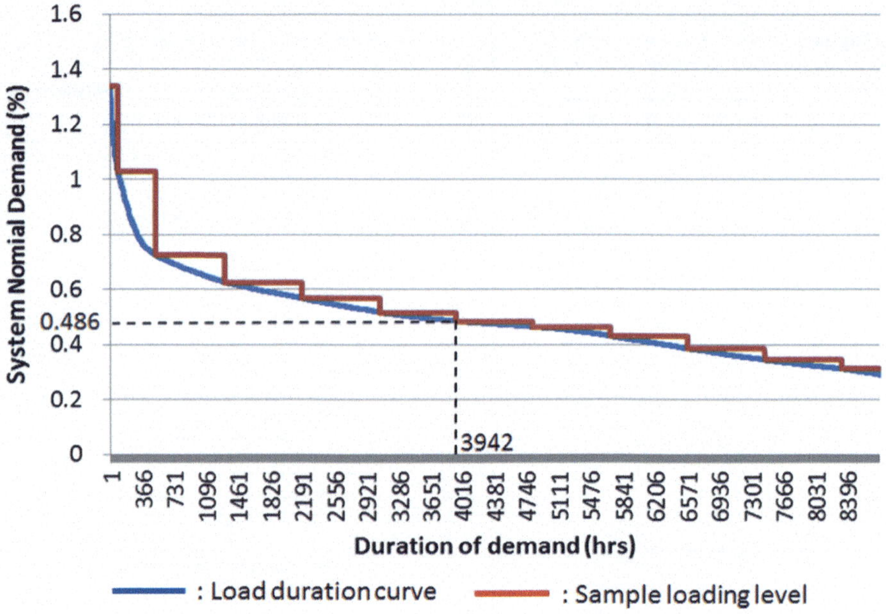

Fig. 6.11 The annual loading curve

discrete load duration curve representing the annual loading conditions of NETS-NYPS test system, which is illustrated in Fig. 6.11. These data are collected from real power system for duration of one year.

In Fig. 6.11, the horizontal axis is the duration of demand measured in hours. There are 8760 h in total in a year. The vertical axis is the percentage of the system nominal demand. The maximum demand corresponds to loading factor of 1.318, which is also the maximum loading used in Chap. 5. The annual loading curve is interpreted as this: for any loading factor λ on the curve, its corresponding duration of demand t means that the loading factor is not less than λ for t hours. For example, for the loading factor of 0.486, the corresponding duration of demand is 3942, which means the system loading factor is not less than 0.486 for 3942 h. The annual loading curve is sampled at 12 different duration of demand as shown in Table 6.5. It is sampled every 10% of the duration of demand, except that at high loading levels, where the curve is very steep, it is sampled at 0, 1, and 6% of the duration of demand.

In the following studies, the ranking of load locations are performed for each of 12 loading conditions. The overall ranking summarises the ranking at different loading conditions by counting the frequency each load location appears in each ranking position.

Table 6.5 The selected 12 loading conditions

Operating condition	Duration of demand (%)	Loading factor
1	0	1.318
2	1	1.027
3	6	0.726
4	15	0.627
5	25	0.568
6	35	0.518
7	45	0.486
8	55	0.465
9	65	0.432
10	75	0.387
11	85	0.346
12	95	0.314

6.5.3 Critical Load Locations for Voltage Stability

The critical load locations are firstly been studied for voltage stability. The ranking results for 35 load buses for rated loading condition is illustrated by using bar chart in Fig. 6.12.

It can be seen from Fig. 6.12 that the ranking is not heavily affected by the loading, which is usually the case in the parameter ranking in Chap. 5. In order to have a better understanding of where these important buses are located at, Fig. 6.13 shows the test network with the most critical load bus locations marked with a red circle.

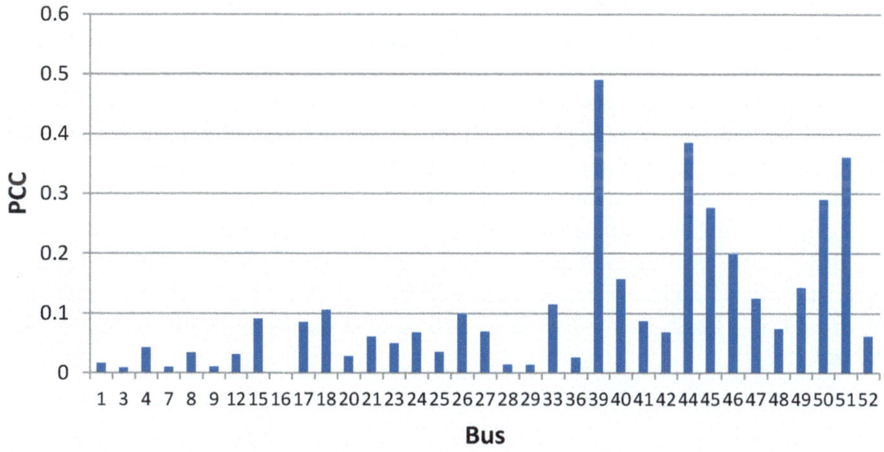

Fig. 6.12 Ranking of load locations for voltage stability

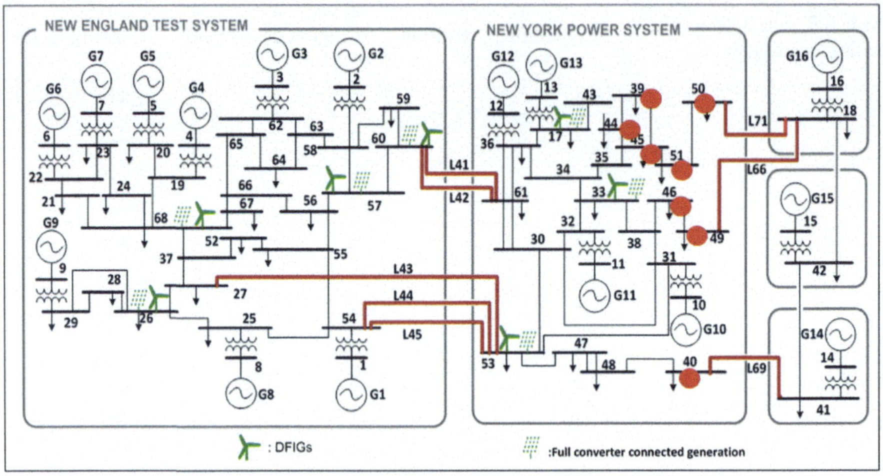

Fig. 6.13 Critical load locations for voltage stability

Figure 6.13 clearly shows that the critical load locations are load buses that are far from generators. The top five most critical locations, Buses 39, 33, 45, 50, 51 are in the same area and connected with each other. These five buses form a loading centre that is far from any generators. In addition, Bus 51 is more critical than Bus 50, and Bus 51 is further away than Bus 50 from their nearest generator G16. Bus 39 is the most important and it has the largest distance to any generators. Bus 46 is more critical than Bus 49, and Bus 46 is farther than Bus 49 to their nearest generator G16. The buses where large loads are located at do not appear in the critical load locations. Bus 17, 18, 41, and 42 are directly connected to generators. The load locations that are far from generators tend to be more critical because for the same amount of load increase, the power from the generator will be more difficult to be transferred to these buses than to the buses that are close to the generator. There are losses in lines and the power could be absorbed by other loads through the transmission line. This lack of power at these far from generator buses will result in voltage drops at these buses, thus deteriorating the voltage stability. Table 6.6 summarises the frequency of each bus appearing in each ranking position for 12 different loading conditions.

The results show that the ranking order is stable. The final ranking order is obtained according to which bus appears most frequently at each ranking position. For example, Bus 44 appears 3 times at the 3rd place, Bus 51 appears 8 times, Bus 50 appears once, and Bus 45 appears once as well, so Bus 51 is chosen to be the bus ranked at the 3rd place. Therefore, the five most critical load locations are Buses 39, 44, 51, 50, and 45.

Table 6.6 The summary of ranking orders for 12 loading conditions for voltage stability

Ranking	Bus number								
	39	44	51	50	45	46	40	49	33
1	10	1	1						
2	2	7	3						
3		3	8	1	1				
4				6	3	2		1	
5				4	5	1	1		1

6.5.4 Critical Load Locations for Small Disturbance Stability

The same analysis has been performed for small disturbance stability. The ranking result at the rated loading condition is illustrated in the Fig. 6.14.

Again, it can be seen that the ranking is not dominated by the load size. In order to have a better understanding of where the important buses are located at, the critical buses are marked with a red dot in the test network diagram in Fig. 6.15.

Comparing Figs. 6.13 and 6.15, it can be seen that the critical load locations have changed. For voltage stability, the critical load locations are load centres and far

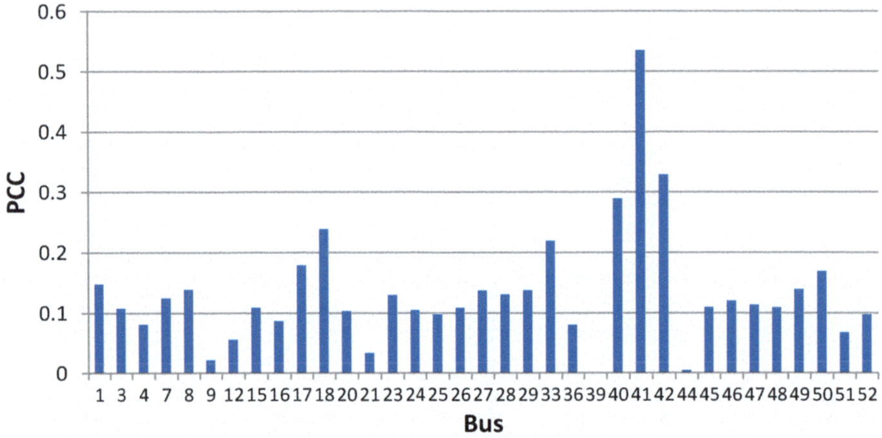

Fig. 6.14 Ranking of load locations for small disturbance stability

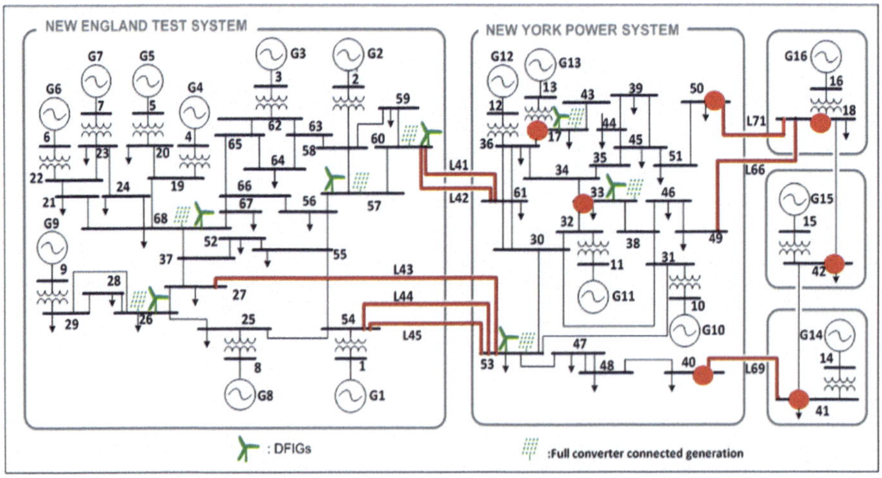

Fig. 6.15 Critical load locations for small disturbance stability

Table 6.7 The summary of ranking orders for 12 loading conditions for small disturbance stability

Ranking	Bus number							
	41	42	40	18	33	17	50	1
1	11	1						
2	1	9	2					
3		2	8	1	1			
4			1	8	2		1	
5			1	2	7	1		1

from generators. However, for small disturbance stability, the critical load locations are buses that are near to generators. The most critical location, Bus 41 is directly connected to generator G14 and connected to G15 though a transmission line. The second critical location, Bus 42 is directly connected to G15 and connected to G16 via a transmission line. Bus 17 is also directly connected to a generator. Bus 40 and 50 are connected to a generator by a tie line. Bus 33 is connected to G11 by a transmission line and there are no other loads between Bus 33 and G11. The buses in New England Test System are less important because the generator sizes in this area (G1–G9) are smaller than G10–G16. The importance ranking of locations can explain why Bus 41 is more critical than Bus 42 and Bus 18 for small disturbance stability even though Bus 42 and Bus 18 are larger than Bus 41. Table 6.7 summarises the frequency of each bus appearing in different ranking position for 12 loading conditions.

From Table 6.7, the five most important load locations are Buses 41, 42, 40, 18, and 33. The ranking order does not have large variation for different loading conditions.

6.5.5 *Critical Load Locations for Transient Stability*

The critical load locations are then identified for transient stability. The ranking result at the rated loading condition is illustrated in Fig. 6.16.

Figure 6.16 shows that the critical load locations for transient stability are similar to the critical load locations for small disturbance stability, i.e. the locations that are close to generators are more important. That is because they are both rotor angle stability. However, there are still some differences between the ranking for small disturbance stability and transient stability. For transient stability, the importance of buses which are close to large generators, i.e. G12, G13, G14, and G16 have increased. The critical load locations are marked with red dots and illustrated in Fig. 6.17. Table 6.8 shows the frequency of each load bus appearing in different ranking position.

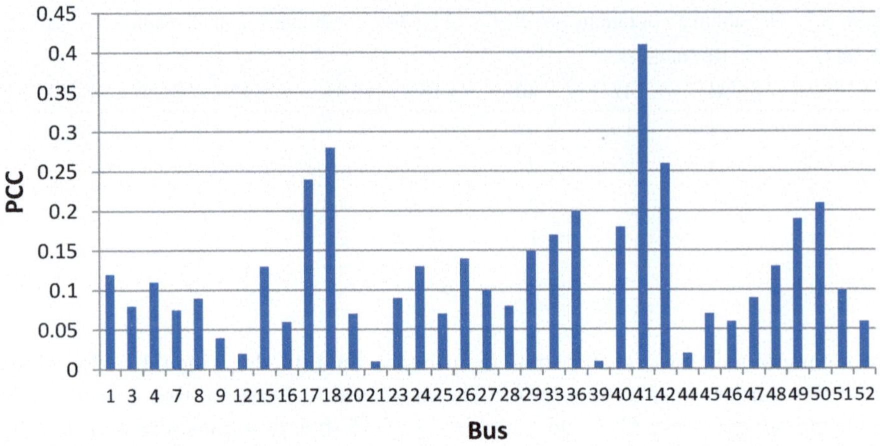

Fig. 6.16 Critical load locations for transient stability

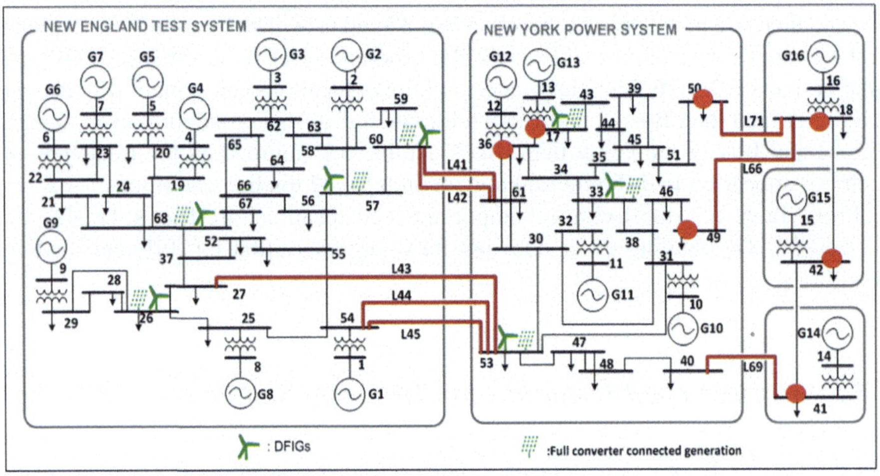

Fig. 6.17 Critical load locations for transient stability

Table 6.8 The summary of ranking orders for 12 loading conditions for transient stability

Ranking	Bus number							
	41	18	42	17	50	36	49	40
1	10	2						
2	2	8	1	1				
3		2	9	1				
4			1	7	2	2		
5			1	2	7		1	1

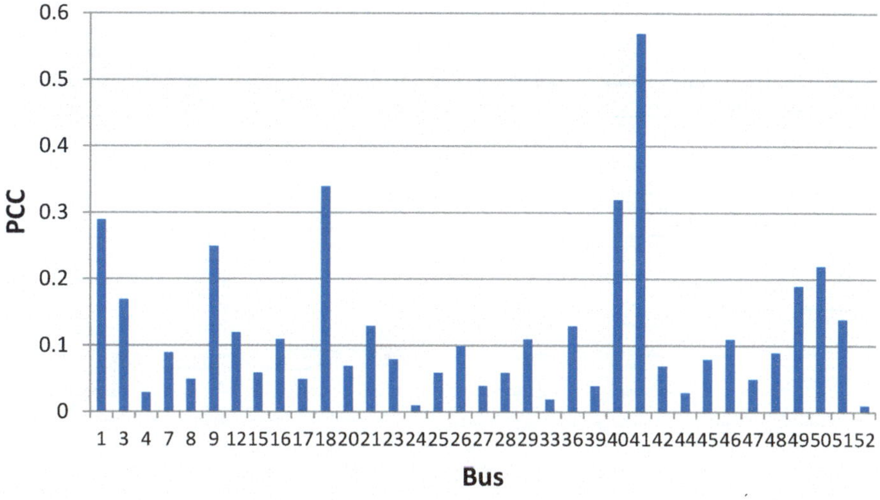

Fig. 6.18 Critical load locations for frequency stability

6.5.6 Critical Load Locations for Frequency Stability

Finally, the important load locations for frequency stability are identified. Similarly, the ranking result for rated loading condition is shown by Fig. 6.18.

Figure 6.18 shows that the critical load locations for frequency stability are buses that are connected with tie lines (Buses 1, 9, 18, 40, 41, 49, 50). The load variation on these locations will have large impact on the power flow, thus heavily affecting the frequency. The location of Bus 41 is very critical, which makes Bus 41 the most critical load bus for frequency stability, though it is not the largest load. Also, as the importance of location of Bus 18 is more influential than Bus 17, Bus 18 has a similar influence on frequency stability with Bus 17 even though load 17 is larger than load 18. The locations of these critical buses are marked with a red dot on the test network diagram as before and are illustrated in Fig. 6.19. Table 6.9 shows the frequency of each bus appearing in different ranking positions for 12 loading conditions.

6.5.7 Summary on the Effects of Load Location

In this section, the power system buses are ranked by the importance of their locations for four types of power system stability studies. The loading of each bus are varied for the same amount. The obtained ranking orders are not heavily affected by the load sizes, thus they are different compared with the ranking orders obtained in Chap. 5. The ranking orders for different types of stability also vary significantly. For voltage stability, the critical load locations are the load centres that are far from generators.

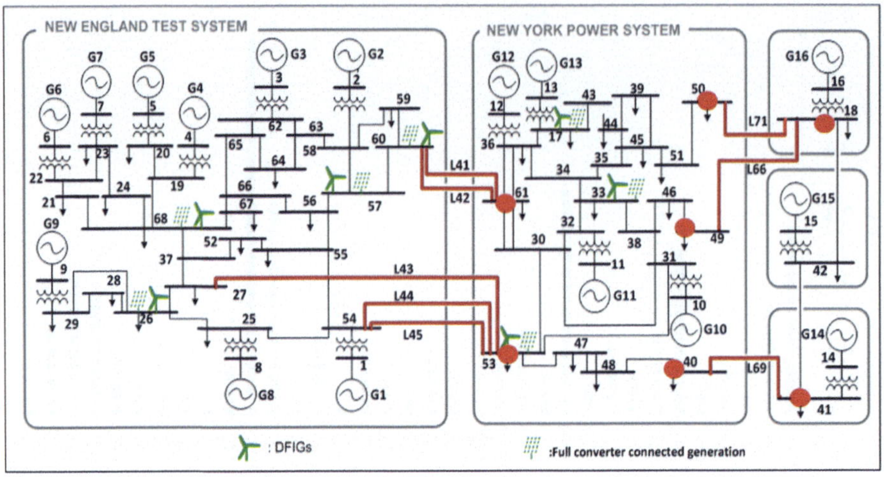

Fig. 6.19 Critical load locations for frequency stability

Table 6.9 The summary of ranking orders for 12 loading conditions for frequency stability

Ranking	Bus number							
	41	18	40	1	9	50	49	3
1	12							
2		10	1	1				
3		2	9	1				
4			2	7	2	1		
5				3	7		1	1

For small disturbance stability, the important load locations are the buses that are close to generators. For transient stability, the influential load locations are also buses close to generators, but the buses that are closer to large generators tend to be more critical. And for frequency stability, the important locations are the buses connected with tie lines between areas. Figure 6.20 illustrates the critical locations for four types of power system stability.

In Fig. 6.20, it can be seen that Buses 18, 40, 41, and 49 are critical load locations for three types of stability. Bus 50 is critical location for all four types of stability.

6.6 Summary

This Chapter investigated how different factors affect the ranking of power system loads and their parameters. Four major factors have been looked into, which are load size, stochastic dependence of load model parameters, load model type, and load location.

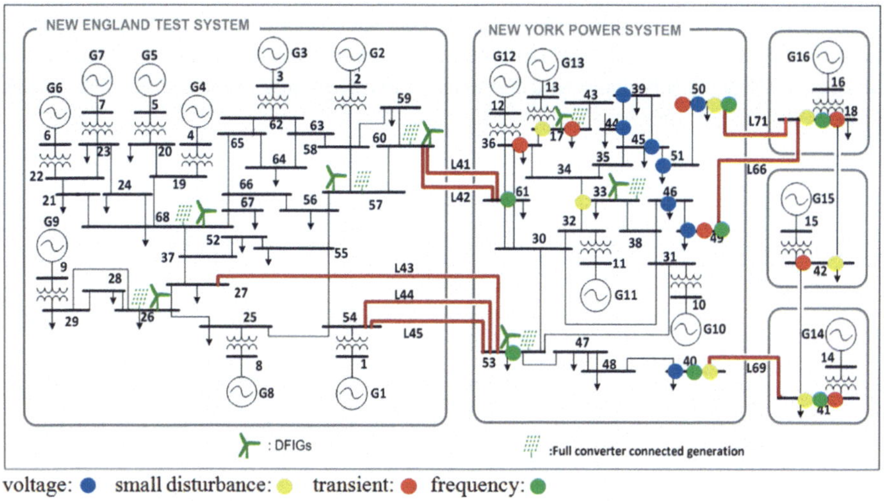

Fig. 6.20 Critical load locations for four types of stability

Rankings in Chap. 5 showed that load size parameter is often the most critical load model parameter. In this chapter, the relationship between the load size and effect of load is reflected in scatter plots using the ranking results obtained in Chap. 5. The results show that the influence of load size is more obvious for large loads. There is nearly a linear relationship between the load effects and logarithm of load size of loads larger than 330 MW. The load would be more important if it is much larger than other loads.

Power system load model parameters are usually correlated with each other, therefore this chapter also looked into the effect of this correlation on load model parameter ranking. The results show that when two parameters are correlated with each other, their ranking position will be closer compared with the ranking position obtained without correlation. This shows that the correlation between load model parameters should be taken into consideration when possible in order to improve the accuracy of load model parameter ranking.

Then the effect of load model type on the load ranking has been studied. In this study, only the load model of one load is changed from static exponential load model to polynomial and composite load model while the load model of other loads remained the same. The results show that when the load model is changed from static load model to dynamic load model, the load would become more influential. The results also show that the ranking of less important loads is more sensitive to change of load model than the ranking of the most important loads.

Finally, this chapter identified the critical load locations in the power system. In order to decouple the load location and the load size, a different ranking method is used, which is based on the Pearson correlation between the variation of loading on each load location and the variation of stability indices. The ranking results show that for different types of stability, the critical load locations are different. For voltage stability, they are load centres far from generators. For transient and small disturbance stability, they are loads close to generators. For frequency stability, they are loads connected with tie lines.

Chapter 7
Required Accuracy Level of Critical Load Model Parameters

7.1 Introduction

Once the critical load model parameters have been identified, the power system operators can focus on building accurate models for those parameters only. However, the accuracy levels that these parameters need to be modelled at, still needs to be determined. This chapter discusses how to determine the accuracy level according to the effect of parameter variations on the confidence level of stability indices.

7.2 Procedure of Obtaining Accuracy Levels

The whole procedure of obtaining accuracy levels of critical load model parameters is illustrated by the flowchart in Fig. 7.1. The procedure contains three stages. Stage I and Stage III are run in MATLAB R2013a, and Stage II is run in DIgSILENT PowerFactory 2017 SP1. The procedure is similar to the procedure of ranking power system load model parameters, but there are still some differences.

The purpose of Stage I is generating uncertainties of load model parameters and renewable generations. The first step is deciding the type of power system stability that is going to be studied. The frequency stability is used as an example here. The next step is choosing the load model, for example, the static exponential load model. Following that one of the critical load model parameters is selected to study its accuracy level requirement. The next step is setting the uncertainty levels for renewable generation. The wind speed follows a Weibull distribution with a scale parameter of 11.1 and shape parameter of 2.2 while the sun light follows a beta distribution with a = 13.7 and b = 1.3 [1]. After that, the uncertainty level of load model parameter is set, for example, $\sigma = 1\%$. For each load model parameter, the simulation will be run for many different uncertainty levels. After all the uncertainty levels have been set, the MATLAB will run the Monte Carlo simulation to obtain uncertain parameters. Then OPF is performed following the retrieving of test network

Y. Zhu, *Power System Loads and Power System Stability*, Springer Theses,
https://doi.org/10.1007/978-3-030-37786-1_7

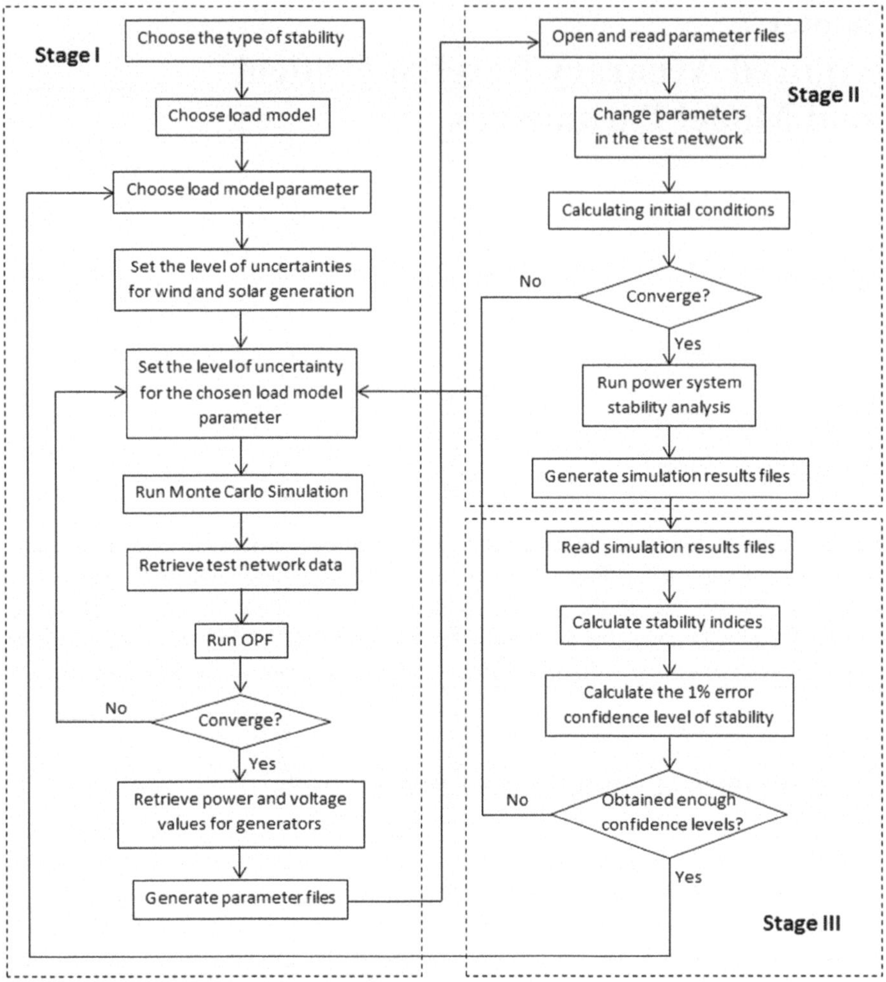

Fig. 7.1 Procedure of obtaining confidence levels

data. If the OPF does not converge, then the uncertainty level of load model parameter needs to be reduced. If OPF is successfully performed, then power and voltage values for generators can be obtained. The final step is generating parameter files for use in Stage II.

The purpose of Stage II is performing power system stability analysis considering uncertainties. The first step is opening and reading the uncertain parameters files from Stage I. Then the values read from these files are used to replace the parameter values in the test network model of DIgSILENT PowerFactory. After that, a power flow is run to calculate initial operating conditions. There is still a possibility that the power flow does not converge because the DIgSILENT PowerFactory uses a more detailed test network model, as described in Sect. 5.1. If this power flow does not converge, then load model parameter uncertainty levels need to be reduced. After the initial conditions have been calculated, the RMS simulation will be run with the same load event for each simulation to find the frequency variation of each bus in the test network after the disturbance. In the final step, the results of each simulation are outputted in a text file for use in Stage III.

The purpose of Stage III is identifying the confidence levels for each uncertainty level of each critical load model parameter. The first step is reading the RMS simulation results from Stage II. Then the obtained frequency values are used to calculate frequency nadirs. After all the frequency nadirs are obtained, the 1% error confidence level of frequency nadir for this uncertainty level is calculated. If sufficient confidence levels are achieved, then a new critical parameter will be chosen. 'Sufficient' means that achieved confidence levels can evenly cover 80–99%. The confidence level lower than 80% usually does not have much practical value. If sufficient confidence levels have not been reached, then the uncertainty level of the parameter will be adjusted according to the new confidence level and repeat all the process after it.

7.3 Comparison of Accuracy Levels for Different Load Model Parameters

7.3.1 Same Parameters for Different Loads

Assuming the load model parameters follow normal distributions, with different standard deviation values σ. By running Monte Carlo Simulation, the confidence level of power system stability index varying within 1% can be obtained. The confidence level will drop for a large σ value, which corresponds to a low accuracy level of parameters. Therefore, by choosing a desired confidence level, the corresponding accuracy level for each load model parameters can be determined. This subsection looks into the accuracy levels of different load model parameters for voltage stability. The critical point of PV curve is thus used as the stability index.

The load model used for all load buses is composite load model. Firstly, the parameter P_0 of Buses 17, 18, and 41 are varied individually, and the corresponding confidence levels of critical points are obtained. The results are shown in Fig. 7.2. The corresponding 3σ values for the 99% confidence levels of each parameter are marked. A 99% confidence level means that we are 99% certain that the variation of critical points will be within 1%. Figure 7.2 clearly shows that as the uncertainty level of load model parameters rises (3σ value increases), the variation of critical points increases. This means that as the accuracy level of load model parameters reduces, the accuracy of the power system stability assessment will drop as well.

From Fig. 7.2, it can be seen that in order to achieve a 99% confidence level, the 3σ values of P_0 of Bus 17, 18, and 41 are 4%, 9%, and 13% respectively. These accuracy levels match their importance ranking in Fig. 5.4, where Bus 17 is the most important, followed by Buses 18 and 41. For more critical parameters, their confidence levels

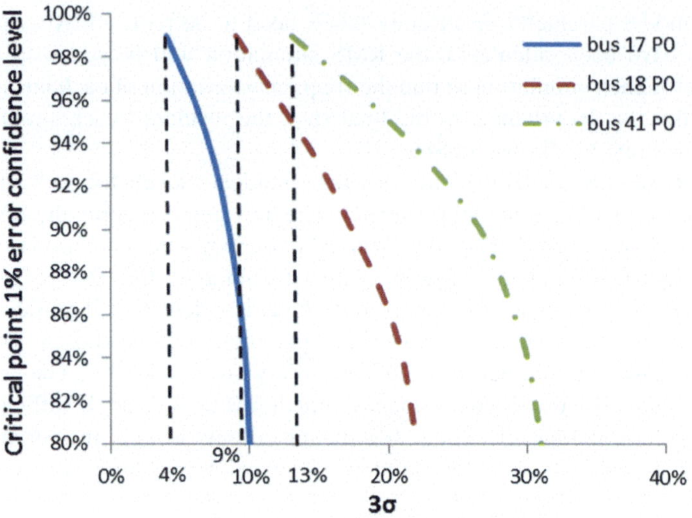

Fig. 7.2 1% error confidence level of P_0 of different buses for voltage stability

also drop more quickly as σ increases. The 3σ value of P_0 of Bus 17 increases from 4 to 9% when confidence level drops from 99 to 90%. In comparison, the 3σ value of Bus 18 increases from 9 to 17% and the 3σ value of Bus 41 increases from 13 to 27%. The confidence level drops faster as σ value increases. Figure 7.3 shows the distribution of nose points when the 3σ value for Bus 17, 18 and 41 is 10%. The influence of load model parameters on nose points variation is more illustrative in

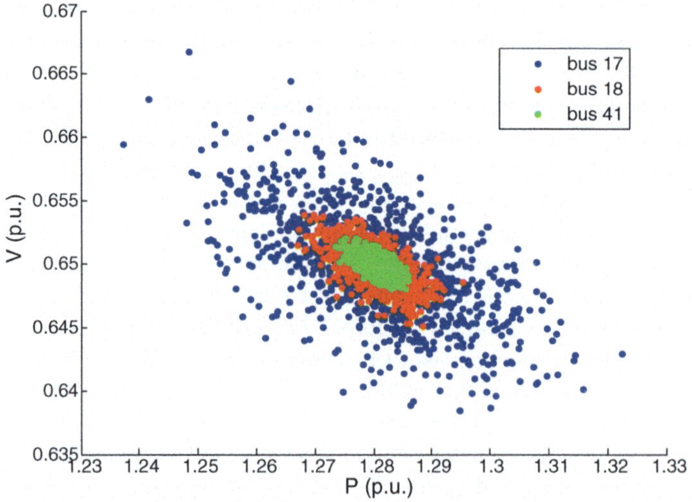

Fig. 7.3 Distribution of PV nose points for the same σ value of different parameters

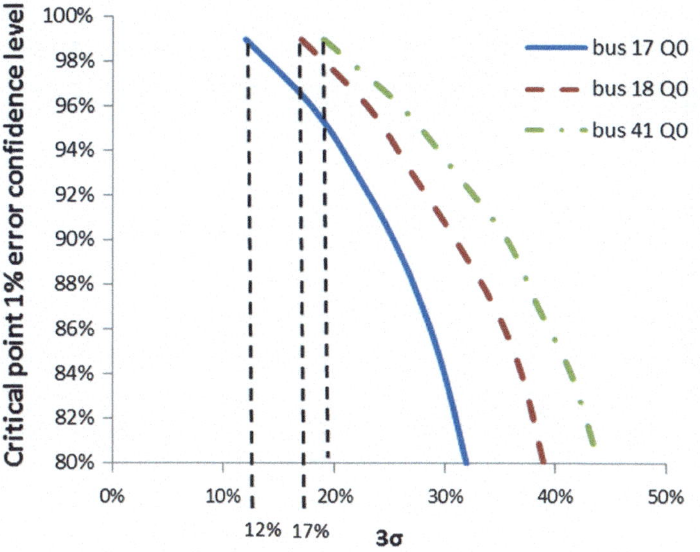

Fig. 7.4 1% error confidence level of Q_0 of different buses for voltage stability

this figure. The blue points, which represent the nose point variation caused by the variation of Bus 17 P_0 are more dispersed than red and green points, which stand for the nose point variation due to Bus 18 P_0 and Bus 41 P_0.

As a comparison, the confidence levels of parameter Q_0 of the same buses are shown in Fig. 7.4.

Figure 7.4 shows that in order to ensure a 99% confidence level, the 3σ values of Q_0 of Bus 17, 18, and 41 need to be 12%, 17%, and 19%, respectively. These accuracy levels also match their ranking in Fig. 5.4. Obviously, these values are larger than the corresponding accuracy levels of P_0, which corresponds to the fact that P_0 is more critical than Q_0. For a 90% confidence level, the 3σ values of Q_0 of Bus 17, 18, and 41 are 25%, 31%, and 36% respectively. The confidence level decreases more rapidly as σ value grows, but the confidence levels drop slower compared with P_0 in Fig. 7.2.

After two load size parameters, the confidence levels of a dynamic load model parameter T_j of Bus 17, 18 and 41 is compared in Fig. 7.5.

Figure 7.5 shows that in order to achieve a 99% confidence level, the 3σ values for T_j of Bus 17, 18, and 41 are 18%, 21%, and 25% respectively. For a 90% confidence level, the 3σ values are 35%, 39%, and 44% respectively. These results match the ranking obtained in Sect. 5.3. These values are larger than the corresponding values of P_0 and Q_0 because load size parameters are more critical than T_j. The confidence levels first drop more quickly as the σ increases, at around 95% confidence level, then becoming slower when the confidence level drop below 86%.

Figure 7.6 shows confidence levels for one of static load model parameters p_1 of Bus 17, 18, and 41.

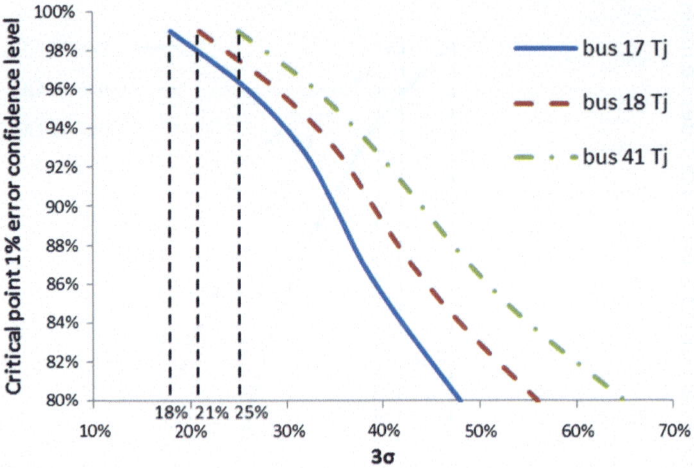

Fig. 7.5 1% error confidence level of T_j of different buses for voltage stability

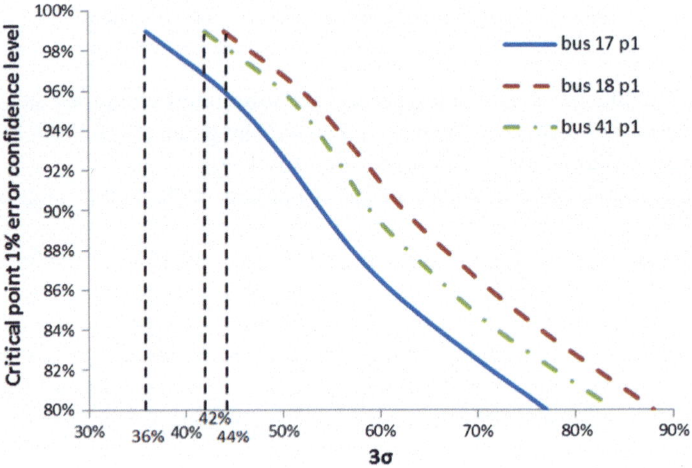

Fig. 7.6 1% error confidence level of p_1 of different buses for voltage stability

As seen from Fig. 7.6, in order to achieve a 99% confidence level, the 3σ values for p_1 of Bus 17, 18, and 41 are 36%, 42%, and 44%, respectively. For 90% confidence level, the 3σ values are 54%, 59%, and 63%, respectively. For P_0, Q_0, and T_j, Bus 18 is always more critical than bus 41. However, for p_1, Bus 41 is slightly more critical than Bus 18. This also matches the ranking results in Sect. 5.3. The 3σ values for p_1 are much larger than the 3σ values of P_0, Q_0, and T_j. This is because the influence of p_1 is much less compared with these three parameters, which is more clearly illustrated in Fig. 7.7 where all composite load model parameters are compared with each other.

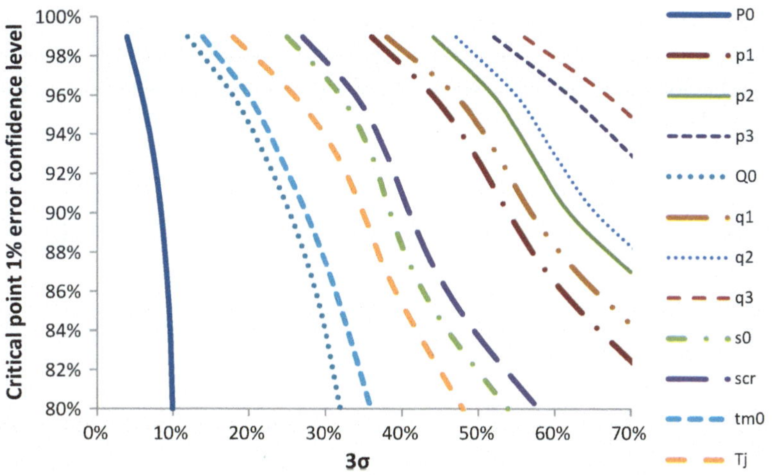

Fig. 7.7 1% error confidence level of all parameters of Bus 17 for voltage stability

7.3.2 Different Parameters of the Same Load

To make this study more comprehensive, Fig. 7.7 compares the different parameters of the same bus. 12 parameters of Bus 17 have been compared. Table 7.1 shows the accuracy levels required for 99% confidence level of each parameter. The 99% confidence level means that we are 99% certain that the error in critical point will be less than 1%.

Table 7.1 and Fig. 7.7 show that there is large variation in the accuracy level required for different parameters. In order to achieve a 99% confidence level, the 3σ value for P_0 is 4% while it is 56% for q_3. Table 7.1 and Fig. 7.7 prove that the less important parameters do not need to be modelled to a high accuracy, as the power system stability studies will not be affected. According to the distance between each line in Fig. 7.7, parameters can be separated into three groups. The first group contains only one parameter, P_0. Its 99% confidence accuracy is 4%, only one third of the second most critical parameters. More efforts are needed when modelling P_0 in order to guarantee its accuracy level. The second group is Q_0, t_{m0}, T_j, s_0, s_{cr}, including one load size parameter and four induction machine parameters. Their 99% confidence accuracies are between 12 and 27%. Care must also be taken when modelling these

Table 7.1 Accuracy levels for 99% confidence level

Parameter	P_0	p_1	p_2	p_3	Q_0	q_1
3σ value (%)	4	36	44	52	12	38
Parameter	q_2	q_3	s_0	s_{cr}	t_{m0}	T_j
3σ value (%)	47	56	25	27	14	18

parameters. The third group contains parameters p_1, p_2, p_3, q_1, q_2, and q_3, which are the proportional parameters of ZIP load model. Their 99% confidence accuracy levels are between 36 and 56%. These are the least important parameters, and it is easy to satisfy their accuracy levels.

7.4 Comparison of Accuracy Levels for Different Stability Studies

7.4.1 *Fixed Standard Deviation*

Section 7.2 compares the accuracy levels for different load model parameters, and the stability index used is the PV curve critical point. This section compares the accuracy levels for different types of stability studies. The parameter P_0 of Bus 17, 18, and 41 is still used as an example, since it is the most critical parameter. Firstly, in order to intuitionally show the effects of these three parameters on stabilities indices, the 3σ value is fixed. They are 3% for transient stability and small disturbance stability, and 30% for frequency stability. The 3σ value for frequency stability is large because for frequency nadir, a 1% variation is very big, which can only results from a large uncertainty. The scatter plot of critical modes is shown in Fig. 7.8, and the histograms of transient stability indices and frequency nadirs are shown in Figs. 7.9 and 7.10. The results are obtained by running Monte Carlo simulation in MATLAB

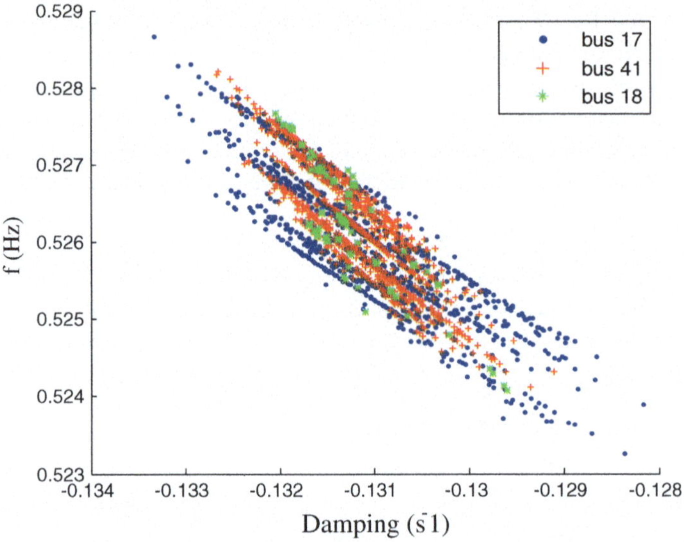

Fig. 7.8 Distribution of critical modes for the same σ value for different parameters

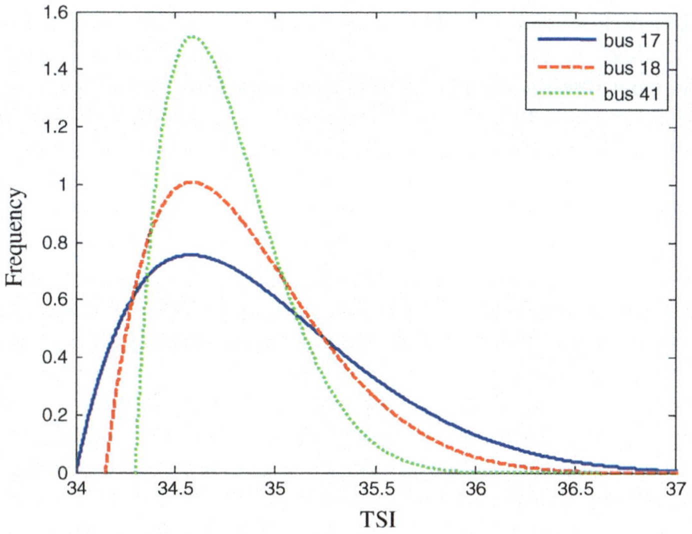

Fig. 7.9 Distribution of TSI values for the same σ value for different parameters

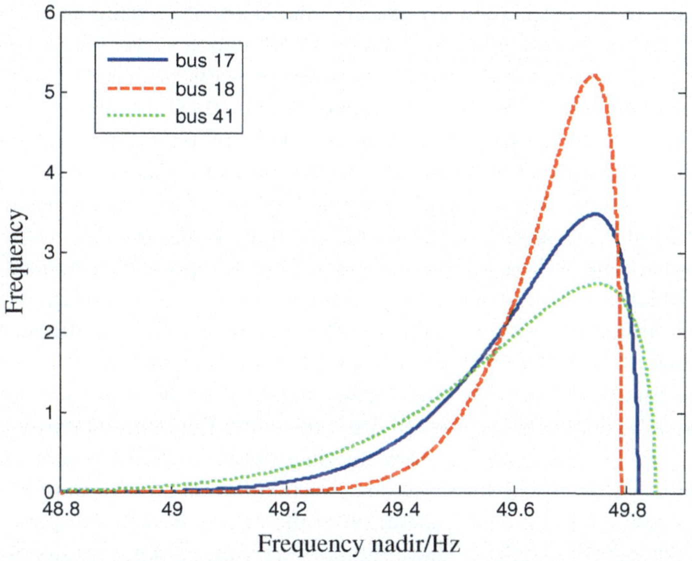

Fig. 7.10 Distribution of frequency nadir for the same σ value for different parameters

and DIgSILENT PowerFactory. Thus they validate the results obtained by Morris Screening method. In Fig. 7.8, the point markers represent Bus 17, the asterisk markers represent Bus 18, and plus markers represent Bus 41. In Figs. 7.9 and 7.10, the solid line stands for Bus 17, the dashed line stands for Bus 18 and dotted line stands for Bus 41. The vertical axis label f of Fig. 7.8 refers to the frequency of power system oscillation modes, while the vertical axis label Frequency of Figs. 7.9 and 7.10 represents the frequency of pdf of TSI and frequency nadir, respectively.

From Fig. 7.8 it can be seen that the distribution of critical modes due to the variation of Bus 17 P_0 is most dispersed while the distribution due to the variation of Bus 18 P_0 is most concentrated. This result agrees with the parameter ranking for small disturbance stability in Sect. 5.4. The centre point, i.e. the critical mode when all parameters are at their mean value is $(-0.1297, 0.5259)$. The distribution patterns of three parameters are not identical. Distribution of Bus 17 P_0 has more variation in the damping axis, and distribution of Bus 18 P_0 has more variation in the frequency axis. According to the distribution of critical modes, the system is stable with a σ value of 1%. The most critical damping is around -0.128 with 1000 simulations.

Figure 7.9 presents the pdf of TSI due to the variation of P_0 of three critical buses. The pdf is fitted from the distribution of TSI values. The figure verifies the ranking results obtained in Sect. 5.3 as the blue line has the largest variation in TSI value and the green line has the smallest variation. The distribution patterns of three buses are similar. The most probable TSI value is 34.59, which is the TSI value when all parameters are at their normal values. The system is stable with a 1% variation of P_0 of any buses. The lowest TSI value is around 34 with 1000 simulations.

Figure 7.10 illustrates the pdf of frequency nadir by the variation of P_0 of three critical buses. The pdf is fitted from the distribution of frequency nadir values. As before, they coincide with the results obtained in Sect. 5.5. The green line has the largest variation in frequency nadir and the red line has the smallest variation. The distribution patterns of three buses are similar. The most probable frequency nadir is 49.74 Hz, which is the value when all parameters are fixed. The system is in danger of frequency instability with a σ value of 10%. From Fig. 7.10, the frequency nadir can be as low as 48.8 Hz, 49.0 Hz, and 49.2 Hz for the variation of P_0 of Bus 41, Bus 17, and Bus 18 respectively. This means that the load model at these three buses should have a higher accuracy level in order to ensure frequency stability.

Figures 7.8, 7.9, 7.10 and 7.2 show the distribution of stability indices for four types of stability studies with the same variation of different critical parameters. These four figures provide intuitional illustration of the effects of different parameters on power system stability studies. These figures verify the ranking results obtained in Sect. 7.5. For more critical parameters, the stability indices will have a larger variation than for the other parameters. These figures also show that for the same variation of load model parameters, the influence on different types of stability varies. Some types of stability are more sensitivity to the variation of load model parameters, while some other types are less sensitive.

7.4.2 Varying Standard Deviation

The confidence level of voltage stability index against 3σ value is already shown in Fig. 7.2. Figures 7.11, 7.12 and 7.13 illustrate the variation of confidence levels as 3σ value changes for transient, small disturbance, and frequency stability respectively. The stability indices used are TSI, critical mode damping, and frequency nadir.

Figures 7.2, 7.11, 7.12 and 7.13 show that for different stability studies, the accuracy level requirements are different. Table 7.2 summarises the 3σ value required in

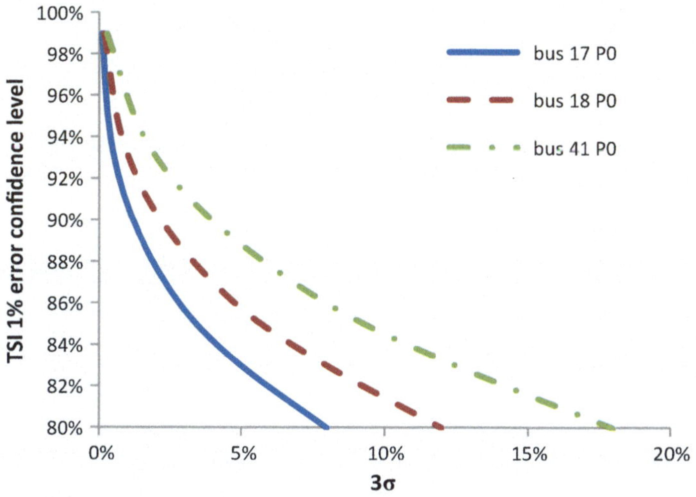

Fig. 7.11 1% error confidence level of P_0 of different buses for transient stability

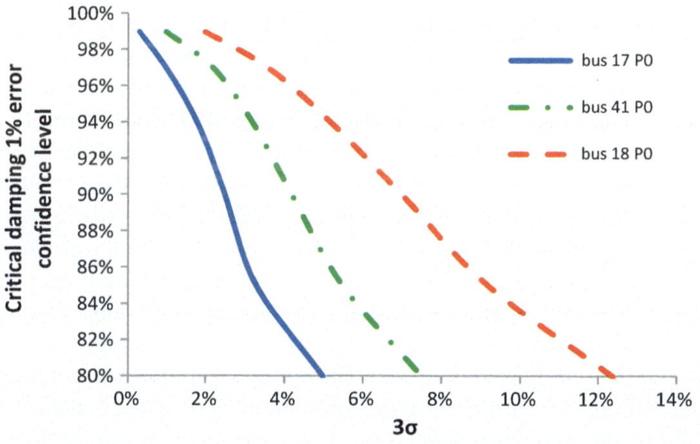

Fig. 7.12 1% error confidence level of P_0 of different buses for small disturbance stability

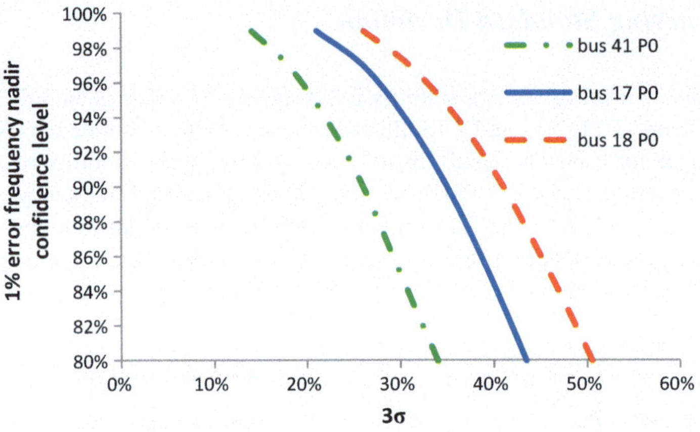

Fig. 7.13 1% error confidence level of P_0 of different buses for frequency stability

Table 7.2 3σ values for different stability studies

Stability	Voltage (%)	Transient (%)	Small disturbance (%)	Frequency (%)
Bus 17 P_0	4	0.05	0.3	14
Bus 18 P_0	9	0.08	2	21
Bus 41 P_0	13	0.1	1	26

order to achieve the 99% confidence level for four different types of stability. From Table 7.2, it can be seen that for the same 1% error confidence level, transient stability requires the highest accuracy level for the same parameter compared with other three stability studies. That is to say, that in order to make sure that we are 99% certain that the error in stability indices will be within 1%, the required accuracy level for the same parameter will be the highest for transient stability. In contrast, the frequency stability has the lowest accuracy level requirement. However, this does not mean that the transient stability is more important than frequency stability, because the margins for their stability index are different. In order to maintain transient stability, TSI only needs to be positive, however, frequency nadir normally should not be 1% lower than the normal frequency [2]. For example, if the TSI value is 60, then 1% error would mean that the TSI varies between 59.4 and 60.6, which will not be a big problem. If the frequency nadir is 49.7 Hz, then 1% error would result the frequency nadir varies between 49.2 and 50.2 Hz. The lower value may result in frequency instability. Therefore, power system operators need to adjust the allowed range of variation for different stability indices.

Figures 7.2, 7.11, 7.12 and 7.13 also show that the trends for confidence level variation are different for different types of stability. For voltage stability and frequency stability, the confidence level drop slowly for small 3σ values, then the slope increases for large 3σ values. For small disturbance stability, the confidence level reduces slowly at first, the speed increases as 3σ values grow larger, but the slope reduces again when 3σ values continue to increase. Finally, for transient stability, the confidence level drops quickly at the beginning. The slope gradually reduces as 3σ values increase. The accuracy levels required for different parameters are consistent with their ranking orders.

7.5 Comparison of Accuracy Levels for Different Loading Conditions

The confidence levels of Bus 17 P_0 have been obtained for maximum, average, and minimum loading conditions for small disturbance stability. They are used to compare with the confidence levels under the rated loading conditions. Because for different loading conditions, the required accuracy levels for load model parameters may vary. The results are illustrated in Fig. 7.14.

Figure 7.14 clearly shows that as the loading reduces, the confidence level for the same parameter variation improves. For instance, when 3σ value equals 2, the confidence levels are 90%, 93%, 94%, and 95% for maximum, rated, average, and minimum loading respectively. Here a 90% confidence level means that we are 90% certain that the error in damping of critical mode will be less than 1%. This difference of confidence level increases as the 3σ value of Bus 17 P_0 increases. That is because at a lower loading condition, the load size parameter value at each bus is also smaller. The load size parameters act as a scale of the output of load models, i.e. the load power. Therefore, the load model parameters will have less impact on power system

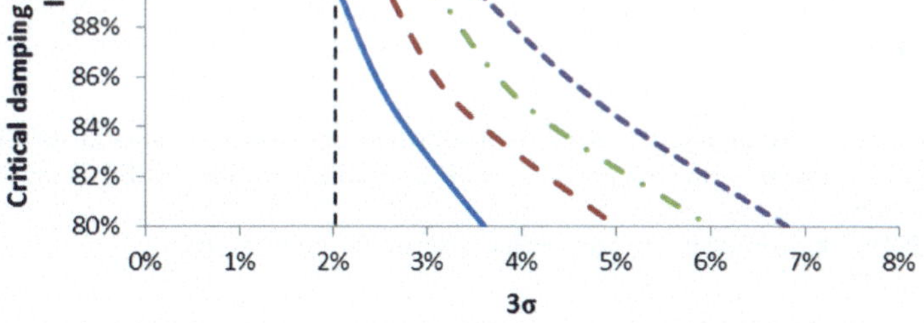

Fig. 7.14 1% error confidence level of Bus 17 P_0 of different loading conditions for small disturbance stability

Table 7.3 Variation of stability indices for bus 17 P_0 and bus 12 P_0

Stability	Voltage (%)	Transient (%)	Small disturbance (%)	Frequency (%)
Bus 17 P_0	3.16	15.5	4.82	0.76
Bus 12 P_0	0.18	0.87	0.25	0.08

stability when the loading level reduces. This also accord with the fact that the power system is more prone to instability when it is heavily loaded.

7.6 Summary

This chapter established the required accuracy level of load model parameters for assessing accurately system stability for different types of stability studies. The confidence levels of 1% error of stability indices are calculated based on Monte Carlo simulation. The accuracy levels for different load model parameters, different stability studies and different loading conditions have been compared. The obtained accuracy levels justify the ranking obtained in Chap. 5, i.e., the more critical load model parameters require a higher accuracy level. Table 7.3 shows the variation of stability indices for the same amount of uncertainty of Bus 17 P_0 and P_0 of one of the least important load buses, Bus 12. The uncertainty level for both of them is $3\sigma = 10\%$. Table 7.3 clearly shows that the variation of stability indices caused by uncertainty in Bus 17 P_0 is significantly larger than that caused by the uncertainty in Bus 12 P_0.

The results in Sect. 7.4 show that for the same confidence level, the required accuracy levels are different for different types of stability. The transient stability requires the highest parameter accuracy, and the variation of transient stability index can also be larger than variation of indices used to assess other types of stability. The results of Sect. 7.5 show that the required accuracy level will reduce for lower loading conditions, i.e., less accurate parameters can be used.

References

1. Papadopoulos PN, Milanović JV (2017) Probabilistic framework for transient stability assessment of power systems with high penetration of renewable generation. IEEE Trans Power Syst 32:3078–3088
2. Grid N (2018) The grid code. In: European connection conditions (ed)

Chapter 8
Conclusions and Future Work

8.1 Conclusions

The original contributions made in this thesis are:

- Development and improvement of the automated load modelling tool (ALMT).
- Ranking load model parameters by using the Morris screening method.
- Determining the required accuracy levels of load model parameters of different loads by Monte Carlo method.
- Investigating the influence of stochastic dependence of parameters on load model parameter ranking by Gaussian Copula method.
- Identifying the critical load locations in power system for different types of system stability studies by Monte Carlo method.

This thesis first explains the importance of developing sufficiently accurate load models. If load models can accurately capture the load performance during the disturbance, power system operators will have more chance to successfully handle the emergency conditions and can have a better control of the power system, as system behaviour would have been more accurately predicted. Some power system collapse events due to inappropriate load models have been reviewed. A review of six frequently used load models has been made, which includes static exponential, polynomial, linear, static induction motor, exponential dynamic, and composite load model. The static exponential, polynomial, linear and static induction motor load model are static load models. The exponential dynamic and composite load model are dynamic load models. This thesis focuses on static exponential, polynomial, and composite load model as was frequently used load models. Two different methods can be used to model power system loads—component based and measurement based load modelling approach. The component based load modelling approach is a bottom-up approach, while the measurement based load modelling approach is a top-down approach. No matter which method is used, load modelling is a complex procedure, which requires significant financial and human resources.

Y. Zhu, *Power System Loads and Power System Stability*, Springer Theses,
https://doi.org/10.1007/978-3-030-37786-1_8

To solve this problem, an automatic load modelling tool (ALMT) has been fully developed and fine tuned in this research. The ALMT, development started originally as a part of the author's undergraduate project. It can automatically build load models from recorded power system data without human intervention. It can achieve a high accuracy and is suitable for online application. The ALMT contains three stages—data processing, load model selection, and load model parameter fitting. In the first stage, the recorded power system data are imported into the software, and useful responses that can be used for building load models are selected. Then these responses are filtered by SG filter and modified for the next stage to use. In the second stage, a load model is chosen according to the shape of the response. If there are multiple responses in a time slot, then the load model will be chosen based on the dominant response. In the third stage, the load model parameters are fitted using least squares method, and are used to produce the simulated responses. The simulated responses are then compared with the recorded responses. If they can match each other, then it means that the developed load models are accurate. If the difference between them exceeds a threshold, the software will roll back to the load model selection stage and choose another load model. The effectiveness of ALMT has been tested on three different case studies. The new contribution made during PhD are adjusting parameters of the overall load modelling process to increase the accuracy of results and implementing composite load model as a standard option in the ALMT. *The final development and improvement of ALMT is the first original contribution of this thesis.*

Although the load models play an important role in the power system stability studies, it is not necessary and practical to model all loads, or to estimate all their parameters very accurately, as the influence of different loads and load model parameters on power system stability studies various a lot. Some load model parameters have critical effects, while the influence of some other load model parameters can be ignored. Therefore, it is crucial to identify which load model parameters need to be modelled accurately, and which load model parameters require lower accurate level. Because the load model parameter ranking will be performed based on their influence on power system stabilities, four types of power system stability have been reviewed, including voltage stability, frequency stability, small disturbance stability, and transient stability. An introduction has been made about the corresponding stability indices used in the thesis, and short overview of the most widely used ones is presented.

A review of past work in the area of ranking power system loads has also been made in Chap. 1. The past ranking of power system loads has only been done for voltage and small disturbance stability. For transient stability, the ranking has only been done for generators. For frequency stability, no such ranking has been performed for any power system components. In addition, for the previous power system load ranking, only the critical loads are identified, not the load model parameters. In order to have a comprehensive understanding of the important loads in a power system, this research provides a framework of ranking power system load model parameters for different types of power system stability studies. The framework is based on the Morris screening method, thus a review of different sensitivity analysis techniques, especially the Morris screening method has been made in Chap. 4. This review compares three types of sensitivity analysis techniques—local sensitivity method, global sensitivity method, and screening sensitivity method, and shows the advantages of screening sensitivity method.

The framework for ranking load model parameters contains three stages. The first stage is generating load model parameters using Morris screening method, which is performed in MATLAB. In this stage, different load model and loading conditions are chosen. The second stage is performing power system stability analysis using the parameters obtained in Stage I, which is performed in DIgSILENT PowerFactory. The third stage is calculating the stability indices from the simulation results and ranking load model parameters. The ranking is based on the sensitivity of power system stability indices to the variation of each load model parameter. The ranking is performed for four types of power system stability, three types of load models, and four different loading conditions. The ranking results show that (i) for the critical load buses, all its parameters tend to be more critical than the same parameters on other load buses; (ii) active power parameters are usually more critical than reactive power parameters; (iii) the set of critical load buses generally remain unchanged for different stability studies and load models; (iv) the large loads are usually the most critical loads for all types of stability studies. *The developed framework for ranking load model parameters for different power system stability studies is the second original contribution of this thesis.*

After solving the problem of identifying the critical load model parameters in power system, the next step is determining how accurately the important load model parameters need to be modelled. The probabilistic assessment was required for this analysis. Thus, a review of probabilistic modelling of power system uncertainties and Monte Carlo simulation has been performed and summarised in Chap. 4. A framework for determining the required accuracy levels of load model parameters is proposed in Chap. 6. This framework also contains three stages. The first stage is generating uncertainties of load model parameters and renewable generations. In this stage, load model parameter that is going to be studied is chosen, and the uncertainty level of renewable generation and chosen load model parameter are set. Then the Monte Carlo simulation is performed in MATLAB. The second stage is performing power system stability analysis considering uncertainties. The third stage is identifying the confidence levels for each uncertainty level of each critical load model parameter. At this stage, the stability indices are calculated by using the simulation results obtained at previous stages and the confidence levels of stability indices are calculated. After that, different uncertainty levels are chosen and the process is repeated in order to obtain a curve of confidence level against uncertainty level. The desired load model parameter accuracy level can thus be found according to the confidence level requirements of power system stability indices. The obtained required accuracy levels of load model parameters correspond to the ranking orders obtained in Chap. 5. The important loads identified in Chap. 5 will have higher required accuracy levels than those less important loads. The results also show that in order to achieve the same confidence level, the transient stability requires the highest accuracy level of load model parameters compared to other three types of stability. *The framework of identifying the required accuracy levels of load model parameters is the third original contribution of this thesis.*

Having in mind that load model parameters are actually correlated, it is necessary to consider the dependence of load model parameters on each other. The independent probability distributions and random sample generation techniques used in simulations do not take correlations into consideration, which may result in significant errors. A review of modelling stochastic dependence of load model parameters has therefore been performed and summarised in Chap. 4. The Gaussian Copula is used in this research to model the stochastic dependence of load model parameters which have been obtained from realistic field measurements. The generated correlated load model parameters are used for the parameter ranking. The newly obtained ranking results are compared with the ranking without the correlation. The comparison shows that if two parameters are correlated with each other, then their influence on the power system stability will be similar. There are large differences between the ranking results with and without correlation, that is, the ranking orders of load model parameters that are correlated with each other will get closer to each other. This shows that it is necessary to take stochastic dependence of load model parameters into consideration when modelling them. *The investigation of stochastic dependence of load model parameters on their influence on power system stability studies is the fourth original contribution of this thesis.*

The load model parameter ranking results presented in Chap. 5 show that load size parameter P_0 is often the most critical parameter. Thus, the effect of load size on the ranking of loads is studied in the beginning of Chap. 6. The research shows that there is almost a linear relationship between the effects of loads (which is the sum of effects of all the parameters of that load) and the logarithm of load size for large loads in the system. This relationship exists for all four types of power system stability, but is most obvious for transient stability. The effect of load model types on the ranking of loads is also studied in Chap. 6. It is found out that if all other conditions are the same, the load modelled as composite load model will have a larger effect on power system stabilities than when it is modelled as static load models.

Considering that the location of load, not only its size, plays important role, the last part of Chap. 6 identifies the critical load locations. They are identified by calculating the Pearson correlation between the load variation of these locations and stability indices. The load location rankings obtained are found to be different from the original load ranking. The most critical load locations were not those where large loads were connected. Also, for different types of power system stability, the load location rankings are very different. The critical load locations for voltage stability are the buses in load centres and far from generators. The important load locations for frequency stability are the ends of tie lines between areas. The load locations that are influential on small disturbance stability are the buses near generators. The critical load locations for transient stability are also buses near generators, but the difference is that the buses near large generators are often more important compared with the locations identified for the small disturbance stability. *Identifying critical load locations for different types of power system stability studies is the fifth original contribution of this thesis.*

8.2 Future Work

The research presented in this thesis has achieved the aim and all the objectives specified in Chap. 1. This last section of the thesis identifies the areas where some future research could be done in this general area.

The first area for future work could be looking into the influence of stochastic dependence of load model parameters on parameter ranking for different types of load models. In this thesis, only the correlations of voltage exponents of static exponential load models are studied principally to illustrate the concept and identify whether this is an area that should be explored. The influence of stochastic dependence of parameters of other load models still needs to be studied, especially for the composite load model. The same process can be followed as in this thesis. The difference is that for composite load model, for example, multiple parameters may be correlated with each other, which will make it difficult to study and more advanced techniques might need to be employed.

The second area for future work could be the development of an analytical method of ranking power system load model parameters. The framework used in this thesis is based on repeated, numerous, simulations. This is time consuming and computationally expensive in particular for large power system. Therefore, if appropriate analytical method can be developed, the process will be less time consuming. This research however may face many difficulties. The first is that for different power system stability the analytical ranking method will probably be different. Thus four analytical methods need to be developed. The second is that the analytical methods usually simplify power system components. The developed method may not be applicable to complex realistic system models. The third is that if the power network model is built in commercial power system analysis software, like DIgSILENT PowerFactory, the analytical method may not have access to all the information it needed, for example, the Jacobian Matrix.

The third area for future work could be identifying influential loads for other aspects of system operation. One area that has been intensively studied is load shedding, i.e., disconnecting the most effective loads when the power system is close to instability. The other area that is becoming particularly important is identifying the loads that could participate in provision of ancillary services. First it would be needed to identify what types of ancillary services are expected to be provided and then load locations which are more suitable for providing this ancillary service than the others would need to be identified. Along those lines, it may also need to be determined how accurately the load model parameters need to be known for provision of adequate level of ancillary service.

The fourth area for future work could be considering location ranking having in mind future high penetration level of electric vehicles. It will be a huge impact on power system stability and other aspect of system operation if many electric vehicles are charged at the same time at a given location. Identifying system vulnerability with respect to EV charging locations would certainly help with network planning and stable and secure operation of the power system.

The fifth area for future work could be considering some future trends in power systems for load model parameter ranking. These trends include the integration of new types of power electronic loads, growing amount of storage systems, application of demand-side management systems etc. These new trends will have significant impacts on power system operation, stability, and reliability. However, it would be too complex to model and analyse the system if all these trends are taken into consideration at the same time. Thus it is suggested to study these trends individually or implement them one by one.

The sixth area for future work could be considering the stochastic dependence between different system demands on power system stability. It is obvious that many power loads have some forms of correlation. For example, the demand of office area is inversely correlated with the demand of residential area. When modelling power system loads, if these correlations are taken into consideration, then obtained results of system stability studies will be more accurate.

Appendix A
Test Network Data

This appendix provides the system data of NETS-NYPS test network used throughout this thesis.

A.1 Load Data

The load values for the test network are presented in Table A.1.

A.2 Generator Data

The generator data for the test network are presented in Table A.2.

The generator model data for the test network are presented in Tables A.3 and A.4.

In Sect. 4.2, the OPF is performed for the test network. This process calculates the minimum cost of generation for the given loading scenario. The cost function is given by the equation below:

$$Cost = c_0 + c_1 P + c_2 P^2 \ \$/hour \tag{A.1}$$

The coefficient values for each generator are given in Table A.5, which are adopted from [1].

A.3 Line Data

The line impedance data and transformer off-nominal turns ratio (ONR) for the test network is presented in Table A.6.

Y. Zhu, *Power System Loads and Power System Stability*, Springer Theses,
https://doi.org/10.1007/978-3-030-37786-1

Table A.1 Load data for the NETS-NYPS test network

Load	*P* (MW)	*Q* (Mvar)	Load	*P* (MW)	*Q* (Mvar)
1	252.7	118.56	28	206	28
3	322	2	29	284	27
4	200	73.6	33	112	0
7	234	84	36	102	−19.46
8	208.8	70.8	39	267	12.6
9	104	125	40	65.63	23.53
12	9	88	41	1000	250
15	320	153	42	1150	250
16	329	32	44	267.55	4.84
17	6000	300	45	208	21
18	2470	123	46	150.7	28.5
20	680	103	47	203.12	32.59
21	274	115	48	241.2	2.2
23	248	85	49	164	29
24	309	−92	50	100	−147
25	224	47	51	337	−122
26	139	17	52	158	30
27	281	76			

Table A.2 Generator data for the NETS-NYPS test network

Gen	*P* (MW)	*Q* (Mvar)	*Q*max (Mvar)	*Q*min (Mvar)	*Vg* (p.u.)	*Base* (MVA)	*P*max (MW)	*P*min (MW)
G1	250	0	280	−210	1.045	100	297.5	29.75
G2	545	0	600	−450	0.98	100	637.5	63.75
G3	650	0	720	−540	0.983	100	765	76.5
G4	632	0	720	−540	0.997	100	765	76.5
G5	505	0	560	−420	1.011	100	595	59.5
G6	700	0	800	−600	1.05	100	850	85
G7	560	0	640	−480	1.063	100	680	68
G8	540	0	600	−450	1.03	100	637.5	63.75
G9	800	0	880	−660	1.025	100	935	93.5
G10	500	0	560	−420	1.01	100	595	59.5
G11	1000	0	1120	−840	1.00	100	1190	119
G12	1350	0	1520	−1140	1.0156	100	1615	161.5
G13	3591	0	3360	−2520	1.011	100	3570	357
G14	1785	0	2000	−1500	1.00	100	2125	212.5
G15	1000	0	1120	−840	1.00	100	1190	119
G16	4000	0	4440	−3330	1.00	100	4717.5	471.5

Table A.3 Generator model data for the NETS-NYPS test network (1)

Gen	X_{lk} (p.u.)	X_d (p.u.)	X_q (p.u.)	X'_d(p.u.)	X'_q(p.u.)	X''_d (p.u.)	X''_q (p.u.)
G1	0.0125	0.1	0.069	0.031	0.028	0.025	0.025
G2	0.035	0.295	0.282	0.0697	0.06	0.05	0.05
G3	0.0304	0.2495	0.237	0.0531	0.05	0.045	0.045
G4	0.0295	0.262	0.258	0.0436	0.04	0.035	0.035
G5	0.027	0.33	0.31	0.066	0.06	0.05	0.05
G6	0.0224	0.254	0.241	0.05	0.045	0.04	0.04
G7	0.0322	0.295	0.292	0.049	0.045	0.04	0.04
G8	0.028	0.29	0.28	0.057	0.05	0.045	0.045
G9	0.0298	0.2106	0.205	0.057	0.05	0.045	0.045
G10	0.0199	0.169	0.115	0.0457	0.045	0.04	0.04
G11	0.0103	0.128	0.123	0.018	0.015	0.012	0.012
G12	0.022	0.101	0.095	0.031	0.028	0.025	0.025
G13	0.003	0.0296	0.0286	0.0055	0.005	0.004	0.004
G14	0.0017	0.018	0.0173	0.00285	0.0025	0.0023	0.0023
G15	0.0017	0.018	0.0173	0.00285	0.0025	0.0023	0.0023
G16	0.0041	0.0356	0.0334	0.0071	0.006	0.0055	0.0055

Table A.4 Generator model data for the NETS-NYPS test network (2)

Gen	T'_{d0} (s)	T'_{q0} (s)	T''_{d0} (s)	T''_{q0} (s)	H (s)	D
G1	10.2	1.5	0.05	0.025	42	4
G2	6.56	1.5	0.05	0.05	30.2	9.75
G3	5.7	1.5	0.05	0.045	35.8	10
G4	5.69	1.5	0.05	0.035	28.6	10
G5	5.4	0.44	0.05	0.05	26	3
G6	7.3	0.4	0.05	0.04	34.8	10
G7	5.66	1.5	0.05	0.04	26.4	8
G8	6.7	0.41	0.05	0.045	24.3	9
G9	4.79	1.96	0.05	0.045	34.5	14
G10	9.37	1.5	0.05	0.04	31	5.56
G11	4.1	1.5	0.05	0.012	28.2	13.6
G12	7.4	1.5	0.05	0.025	92.3	13.5
G13	5.9	1.5	0.05	0.004	248	33
G14	4.1	1.5	0.05	0.0023	300	100
G15	4.1	1.5	0.05	0.0023	300	100
G16	7.8	1.5	0.05	0.0055	225	50

Table A.5 Generator cost data for the NETS-NYPS test network

Gen	c_0	c_1	c_2
G1	0	6.9	0.0193
G2	0	3.7	0.0111
G3	0	2.8	0.0104
G4	0	4.7	0.0088
G5	0	2.8	0.0128
G6	0	3.7	0.0094
G7	0	4.8	0.0099
G8	0	3.6	0.0113
G9	0	3.7	0.0071
G10	0	3.9	0.0090
G11	0	4.0	0.0050
G12	0	2.9	0.0040
G13	0	2.5	0.0019
G14	0	3.3	0.0033
G15	0	3.8	0.0050
G16	0	3.5	0.0014

Table A.6 Line data for the NETS-NYPS test network

From bus	To bus	R (Ω)	X (Ω)	Z (Ω)	Angle (deg)
17	36	0.2645	2.3805	2.3951	83.660
18	49	4.0200	60.3589	60.4926	86.190
18	42	2.1100	31.7400	31.8101	86.197
18	50	0.6348	15.2360	15.2492	87.614
19	68	0.8464	10.3155	10.3501	85.309
20	19	0.3700	7.3000	7.3094	87.098
22	21	0.4232	7.4000	7.4121	86.727
22	23	0.3174	5.0784	5.0883	86.424
23	24	1.1638	18.515	18.5515	86.403
25	54	3.7000	4.5494	5.8640	50.879
26	29	3.0153	33.0625	33.1997	84.789
26	25	1.6928	17.086	17.1697	84.342
27	26	0.7400	7.7763	7.8114	84.564
27	37	0.6877	9.1517	9.1775	85.703
28	29	0.7406	7.9879	8.0222	84.703
28	26	2.2747	25.0746	25.1776	84.816
30	53	0.4232	3.9146	3.9374	83.830

(continued)

Table A.6 (continued)

From bus	To bus	R (Ω)	X (Ω)	Z (Ω)	Angle (deg)
31	53	0.8464	8.6227	8.6641	84.394
31	30	0.6877	9.8923	9.9162	86.023
32	30	1.2696	15.23	15.283	85.235
33	32	0.4232	5.2371	5.2532	85.380
33	38	1.9000	23.4876	23.564	85.375
34	33	0.5819	8.3000	8.3204	85.990
35	34	0.0529	3.9100	3.9104	89.225
36	61	1.1638	10.3684	10.4335	83.596
36	34	1.7457	5.87	6.1241	73.438
38	31	0.58	7.7763	7.7979	85.734
38	46	1.1600	15.0200	15.0647	85.584
39	45	0.0000	44.38	44.38	90.000
39	44	0.0000	21.7400	21.7400	90.000
41	40	3.1740	44.4360	44.5492	85.914
42	41	2.1160	31.7400	31.8105	86.186
43	17	0.2645	14.6	14.6024	88.962
43	44	0.0529	0.5819	0.5843	84.806
44	45	1.3255	38.617	38.640	88.039
45	35	0.3700	9.2575	9.2649	87.711
45	51	2.1160	5.5545	5.9439	69.146
47	48	1.3225	14.1772	14.2388	84.671
48	40	1.0580	11.6380	11.6860	84.806
49	46	0.9500	14.4950	14.5260	86.250
50	51	0.4761	11.6900	11.6997	87.668
52	55	0.5800	7.0357	7.0596	85.287
52	37	0.3700	4.3378	4.3536	85.125
53	47	0.6877	9.9452	9.9689	86.044
53	27	16.9280	169.2800	170.1243	84.289
53	54	1.8500	21.7400	21.8186	85.136
55	54	0.6877	7.9879	8.0174	85.079
56	55	0.6877	11.2677	11.2887	86.507
57	56	0.4200	6.7700	6.7830	86.450
58	57	0.1058	1.3754	1.3795	85.601
59	58	0.3170	4.8668	4.8771	86.273
59	60	0.2110	2.4300	2.4391	85.037
60	57	0.4232	5.8190	5.8344	85.840

(continued)

Table A.6 (continued)

From bus	To bus	R (Ω)	X (Ω)	Z (Ω)	Angle (deg)
60	61	1.2167	19.2000	19.2385	86.374
61	30	1.0050	9.6800	9.7320	84.072
62	63	0.2116	2.2747	2.2845	84.685
62	65	0.2116	2.2700	2.2798	84.675
63	64	0.8460	23.0000	23.0155	87.893
63	58	0.3700	4.3378	4.3536	85.125
65	64	0.8464	23.0000	23.0156	87.892
65	66	0.4761	5.3429	5.3641	84.908
66	56	0.4232	6.8241	6.8372	86.451
66	67	0.9500	11.4793	11.5185	85.269
67	68	0.4760	4.9726	4.9953	84.532
68	21	0.4232	7.1400	7.1525	86.608
68	37	0.3700	4.7000	4.7145	85.499
68	24	0.1587	3.1211	3.1251	87.089

Appendix B
Power Curves

The PV panel power and wind turbine curves are shown in Tables B.1 and B.2.

Table B.1 PV panel power curve

Time (hour)	Solar radiation (W/m^2)	Power (kW)
6	0	119.403
7	0	716.418
8	59.7016	1761.194
9	358.2096	1880.597
10	880.5986	1940.299
11	940.3002	2000.000
12	970.151	1970.149
13	1000.002	1910.448
14	985.0764	1731.343
15	955.2256	1343.284
16	865.6732	1014.925
17	671.643	746.269
18	507.4636	447.761
19	373.135	149.254

Y. Zhu, *Power System Loads and Power System Stability*, Springer Theses,
https://doi.org/10.1007/978-3-030-37786-1

Table B.2 Wind turbine power curve

Speed (m/s)	Power (kW)	Speed (m/s)	Power (kW)
4.169	55.860	12.676	1798.565
5.859	201.611	13.380	1880.236
6.901	397.173	14.197	1939.968
7.775	598.614	15.183	1973.366
8.563	794.228	16.254	1987.753
9.324	995.693	17.239	1987.547
10.056	1197.164	18.620	1993.103
10.845	1401.544	20.000	1989.893
11.690	1597.147	25.296	1991.710

Appendix C
Daily Loading Curves

The daily loadings curves for residential and commercial load are illustrated in Fig. C.1, and the daily loading curves for industrial load are shown in Fig. C.2. The daily loading curves are used in Sect. 2.3.

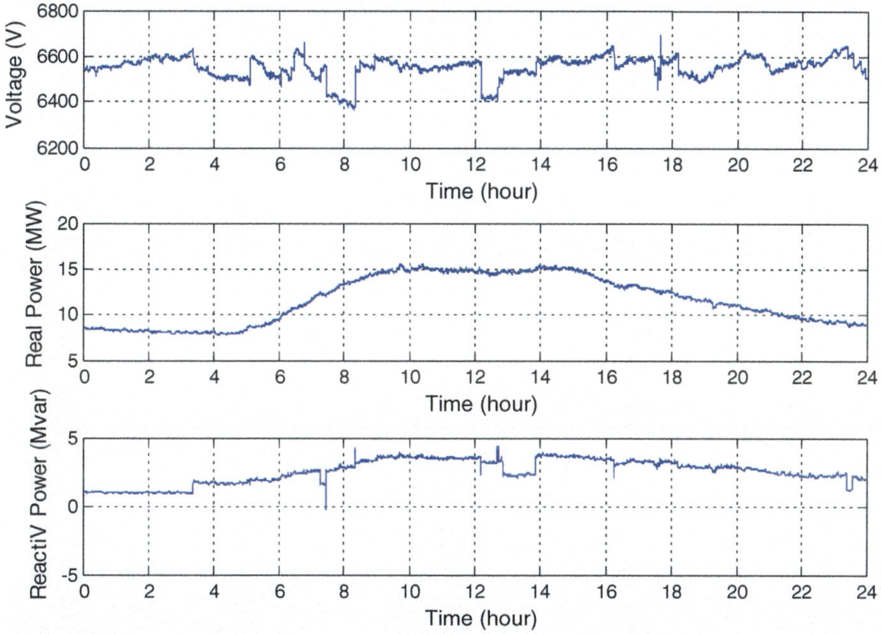

Fig. C.1 A typical daily loading curve for residential and commercial load

Y. Zhu, *Power System Loads and Power System Stability*, Springer Theses,
https://doi.org/10.1007/978-3-030-37786-1

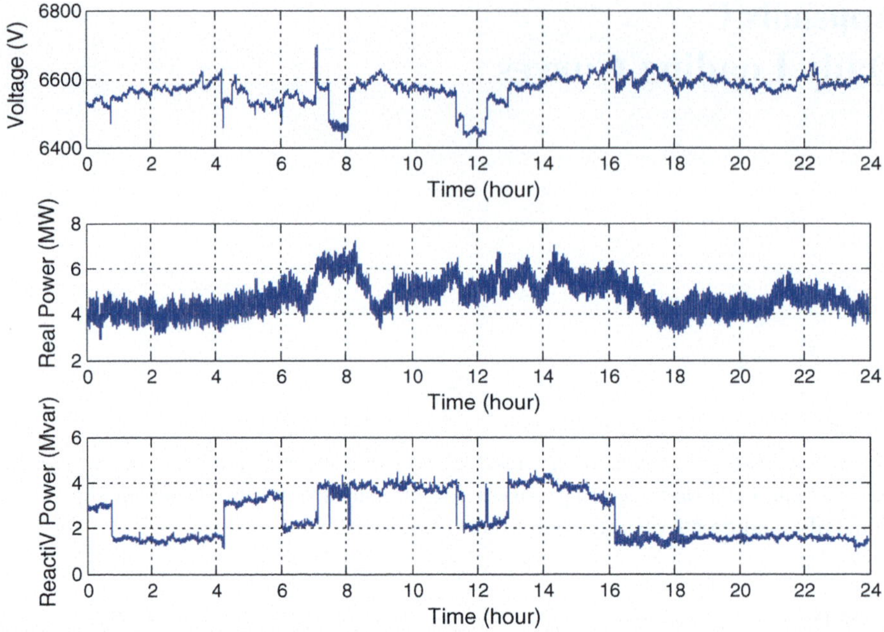

Fig. C.2 A typical daily loading curve for industrial load

Appendix D
Load Model Parameter Ranking Data

The mean of elementary effects of load model parameters of composite load model for voltage stability is presented in Table D.1. The maximum value is modified to 100.

Y. Zhu, *Power System Loads and Power System Stability*, Springer Theses,
https://doi.org/10.1007/978-3-030-37786-1

Table D.1 Composite load model parameter mean of elementary effects for voltage stability

Load	P_0	p_1	p_2	p_3	Q_0	q_1	q_2	q_3	s_0	s_{cr}	T_{m0}	T_j
1	10.11	4.79	0.29	2.60	7.97	4.43	2.70	0.76	5.23	3.06	12.11	9.34
3	15.18	4.26	3.80	0.49	6.77	2.39	1.27	2.03	3.08	3.63	7.65	5.65
4	9.05	3.31	1.78	2.70	15.12	5.13	0.09	1.00	4.42	3.86	7.93	5.57
7	15.10	4.63	0.12	2.13	7.09	4.14	2.16	0.19	5.12	2.71	8.61	6.13
8	8.90	3.71	2.56	1.18	8.21	3.07	0.92	1.71	3.76	4.31	12.16	9.02
9	9.99	4.17	3.26	0.40	6.07	2.69	1.40	2.99	2.80	3.82	7.35	5.42
12	15.16	3.76	2.87	1.35	15.69	3.29	2.53	0.84	3.80	2.73	6.30	4.65
15	28.42	8.36	3.56	2.74	20.42	7.43	4.12	6.02	16.29	8.32	24.66	17.76
16	31.94	7.46	5.24	3.77	15.76	5.74	8.75	2.10	7.88	12.46	24.21	16.89
17	100.00	21.17	10.16	15.67	69.90	15.52	6.70	7.15	32.96	24.54	72.68	50.43
18	79.81	15.75	8.12	7.53	63.32	10.55	8.40	4.41	24.18	16.25	60.02	42.93
20	35.66	10.94	6.16	4.93	45.80	9.58	5.44	3.26	8.75	12.23	36.06	25.97
21	29.67	8.72	3.64	2.42	32.39	7.82	4.32	6.82	16.79	8.86	24.51	17.44
23	9.96	2.44	1.06	2.87	8.63	5.35	1.00	1.22	4.65	3.81	7.59	5.18
24	15.02	4.42	0.18	3.73	7.37	4.84	2.74	0.57	4.97	3.36	12.26	9.33
25	8.22	3.37	2.08	1.64	15.18	3.04	0.73	1.35	3.86	4.38	7.83	5.06
26	15.91	4.80	3.75	0.82	6.38	2.62	1.58	2.53	2.67	3.45	7.65	5.26
27	10.81	3.99	2.03	1.54	7.08	3.81	1.99	0.06	4.14	2.41	6.69	4.99
28	8.53	3.89	2.90	1.63	15.13	2.21	0.18	2.04	3.30	4.62	12.94	8.75
29	32.41	8.99	3.95	2.68	15.99	7.77	4.34	6.66	12.24	8.29	24.68	17.33
33	9.41	2.60	1.75	2.58	8.55	5.58	0.51	1.08	4.72	4.20	7.55	6.02

(continued)

Table D.1 (continued)

Load	P_0	p_1	p_2	p_3	Q_0	q_1	q_2	q_3	s_0	s_{cr}	T_{m0}	T_j
36	10.34	4.89	0.45	3.41	7.21	4.12	1.31	0.73	5.59	3.10	12.00	9.25
39	37.93	18.77	10.04	7.38	44.71	10.73	9.22	5.27	24.67	16.48	48.62	33.83
40	21.17	10.83	6.77	4.94	30.10	9.18	5.10	3.49	8.19	12.90	36.10	25.24
41	64.56	15.78	10.31	9.18	49.34	10.21	7.51	6.91	16.63	24.10	48.39	33.65
42	48.50	9.43	7.00	7.81	27.49	6.90	4.13	7.39	8.93	16.92	55.72	39.12
44	33.34	7.94	5.12	3.73	15.65	5.84	8.40	2.75	8.04	8.32	24.55	17.78
45	9.55	2.33	1.62	2.36	8.76	5.41	0.49	1.70	4.98	3.53	8.04	5.78
46	15.96	4.03	0.36	3.66	15.28	4.23	1.71	0.63	5.39	3.06	8.44	6.23
47	8.45	3.24	2.72	1.86	8.28	3.73	0.13	1.84	3.33	4.59	12.36	9.21
48	15.50	4.49	3.88	0.35	6.45	2.97	1.04	2.98	2.59	3.87	7.79	5.72
49	10.36	3.62	2.81	1.01	7.08	3.98	2.65	0.23	3.60	2.52	6.27	4.46
50	8.33	3.15	2.35	1.12	15.89	2.09	0.93	1.40	3.24	4.34	12.74	9.20
51	42.55	15.69	10.90	8.05	33.30	8.04	6.19	4.72	8.72	24.88	24.58	16.87
52	9.93	4.80	2.28	0.54	6.99	2.72	1.84	2.43	2.87	3.76	7.47	5.79

Appendix E
Publications from the Thesis

E.1 *International Journal Papers*

[E1] Y. Zhu and J. V. Milanović, "Automatic Identification of Power System Load Models Based on Field Measurements," *IEEE Transactions on Power Systems*, vol. 33, pp. 3162–3171, 2018.

[E2] Y. Zhu and J. V. Milanović, "Efficient Identification of Critical Load Model Parameters Affecting Transient Stability," *Electric Power Systems Research*, vol. 175, 2019.

[E3] Y. Zhu, J. V. Milanović, and K. N. Hasan, "Ranking and quantifying the effects of load model parameters on power system stability," *IET Generation, Transmission & Distribution*, vol. 13, pp. 4650–4658, 2019.

E.2 *International Conference Papers*

[E4] Y. Zhu, B. Qi, and J. V. Milanovic, "Probabilistic ranking of power system loads for voltage stability studies in networks with renewable generation," in *IEEE PES Innovative Smart Grid Technologies Conference Europe (ISGT-Europe)*, Ljubljana, Slovenia, 2016.

[E5] B. Qi, Y. Zhu, and J. V. Milanovic, "Probabilistic Ranking of Critical Parameters Affecting Voltage Stability in Network with Renewable Generation," in *IEEE PES Innovative Smart Grid Technologies Conference Europe (ISGT-Europe)*, Ljubljana, Slovenia, 2016.

[E6] Y. Zhu and J. V. Milanović, "Efficient identification of critical load model parameters affecting power system voltage stability," in *IEEE PES PowerTech*, Manchester, UK, 2017.

[E7] Y. Zhu, and J. V. Milanovic, "Identifying Critical Load Locations for Power System Voltage, Angular and Frequency Stability," in *IEEE PES Innovative Smart Grid Technologies Conference Europe (ISGT-Europe)*, Bucharest, Romania, 2019.

Y. Zhu, *Power System Loads and Power System Stability*, Springer Theses,
https://doi.org/10.1007/978-3-030-37786-1

Reference

1. TB Nguyen, M Pai (2003) Dynamic security-constrained rescheduling of power systems using trajectory sensitivities. IEEE Trans Power Syst 18:848–854

The manufacturer's authorised representative in the EU is Springer Nature Customer Service Centre GmbH, Europaplatz 3, 69115 Heidelberg, Germany. If you have any concerns regarding our products, please contact ProductSafety@springernature.com

Printed and bound by CPI Group (UK) Ltd, Croydon, CR0 4YY
30/09/2026
02225458-0002